SKYSTARS

Atheneum   New York   1981

SKY=
STARS

THE HISTORY OF
WOMEN IN AVIATION

by Ann Hodgman & Rudy Djabbaroff

LIBRARY OF CONGRESS CATALOGING IN PUBLICATION DATA

Hodgman, Ann.
  Skystars.

  SUMMARY: Discusses famous and lesser-known women in
aviation from the earliest aviators who flew in balloons in the
eighteenth century to modern flight, including gliders and
helicopters.
    1. Women in aeronautics—Juvenile literature.
2. Women and pilots—Juvenile literature.   [1. Aeronautics—
Biography.   2. Air pilots]   I. Djabbaroff, Rudy.
II. Title.
TL547.H62   629.13′092′2 [B] [920] 81–5075
ISBN 0–689–30870–1   AACR2

Published simultaneously in Canada by
McClelland & Stewart, Ltd.
Manufactured by R. R. Donnelley and Sons Inc.
Harrisonburg, Virginia.
Designed by M. M. Ahern
First Edition

Contents

Illustrations were supplied by the following organizations:

SKYSTARS

The Earliest Flight

IN 1751 a strange and fanciful book was published. It was called *The Life and Adventures of Peter Wilkins,* and it told the tale of a shipwrecked sailor who met and married an enchanting woman named Youwarkee. Youwarkee could fly. As the story went, she lived in a kingdom whose inhabitants all flew: each member of her race possessed a "graundee," a silky skin, which could be spread to create a set of wings. Explaining how it felt to fly, Youwarkee said: "If you had but the graundee, flying would rest you, after the greatest labour . . . once you are upon the graundee at a proper height, all the rest is play, a mere trifle; you need only to think of your way, and incline to it, your graundee directs you as readily as your feet obey you on the ground without thinking of every step you take; it does not require labour, as your boat does, to keep you a-going."

Less than fifty years later came the first actual flight ever made by a woman. This flight took place in one of the most famous inventions of the century: a balloon. Flying in a balloon was not nearly as simple as the mythical flights made by Youwarkee. But the first women who flew in balloons would certainly have agreed with Youwarkee that flight was the perfect way to travel.

The story of the first woman who flew in balloons must start with the story of the first balloons themselves. It all began simply enough. In the eighteenth century, two French brothers, Joseph and Etienne Montgolfier, lived in a town near Lyons called Annonay. Papermakers by trade, they were really more interested in science, and especially in the forces at work in the atmosphere. They had noticed and remembered what many people have noticed and ignored: that bits of ash in a fire are often carried aloft by the smoke. What caused this? the brothers wondered. What made smoke itself rise? They placed little paper bags in the smoke; these miniature balloons also rose gently into the air. Perhaps smoke and hot air could be trapped. If a small fire could lift ash, perhaps a larger one could lift something heavier.

Joseph Montgolfier tried this in 1782, in Avignon. He borrowed some silk from his landlady and sewed it into a bag. Next he set fire to some scraps of paper in his room and held the bag over them. He was delighted, and his patient landlady was astounded, to see the bag swell and float up to bob against the ceiling. "Prepare promptly a supply of taffeta and ropes," Joseph wrote rapturously to Etienne, "and you will see one of the most astonishing sights in the world!" The brothers tried the same experiment outside at Annonay, and the silken bag was lifted almost 100 feet.

By June, 1783, the Montgolfiers had advanced from small silk bags to huge tethered balloons, or, as they called them,

4

"acrostatic machines." They went on to greater fame with the help of a duck, a rooster, and a sheep—the world's unwilling first three air passengers, who lifted off from Versailles in a balloon on September 19, 1783. (The three travelers landed safely in a tree.) A month later, the first human balloon passenger ascended eighty feet in a tethered balloon; this honor belonged to a young man name François Pilâtre de Rozier.

The public became fascinated with ballooning. It had suddenly become possible for humanity to do what it had dreamed about for centuries. Everyone wanted to fly now, and perhaps it was because ballooning was so wildly popular that women began to travel in balloons as early as they did. In fact, the first women to go up in a balloon did so only seven months after François Pilâtre de Rozier had made his own first ascent. Four of them went up at once, in a tethered-balloon exhibition on May 20, 1784. The Marchioness de Montalembert, the Countess de Montalembert, the Countess de Podenas, and Mademoiselle de Lagarde were the daring noblewomen.

Less than a month later, the first woman ever actually to travel in a balloon went aloft from Lyons on June 4—one year after the Montgolfiers had made their first public balloon demonstration at Annonay. She was Madame Thible, another Frenchwoman and a very popular opera singer. A watercolor painted at the time shows Madame Thible and her companion, a painter named Fleurant, riding in state in a dashing pink balloon, with a tremendous crowd to cheer them on. An onlooker remarked that Madame Thible seemed much more courageous than Monsieur Fleurant. She certainly showed no stage fright: at one point in her voyage she entertained the crowds with a comic-opera song called "Oh, To Travel in the Clouds!"

A year later came the voyage of the first English women to fly in a balloon. Like Madame Thible's ascent, Mrs.

Letitia Ann Sage's had a picturesque quality about it. For one thing, Mrs. Sage weighed 200 pounds, a fact not revealed in the flattering painting of her trip. This painting shows Mrs. Sage and her two male companions, Vincent Lunardi and George Biggin, aloft in a blue balloon with a crimson chamber for the passengers. Vincent Lunardi wears the uniform of the British Redcoats and waves a tricornered hat; Mrs. Sage wears a most unsuitable traveling dress and a hat with huge ostrich plumes (which would have blown off instantly in real life), and she reclines on a little fringed sofa.

The painting is flattering. It is also inaccurate. When all three riders were in the balloon chamber, it was discovered that the balloon could not possibly lift such a load. Vincent Lunardi graciously stepped out and let George Biggin and Mrs. Sage ride alone. Their trip was very pleasant. They took off from London with no difficulty; Mrs. Sage did crush the barometer by accident, but otherwise there were no problems on the voyage. The travelers lunched off ham, chicken and wine before landing in a field near Harrow School. A farmer who owned the field tried to destroy the balloon. It had landed on his vegetables. He was unsuccessful, however, Mrs. Sage wrote later, because "the heroic boys of Harrow School saved the balloon from destruction."

Back in France, respectable citizens were beginning to worry about women and ballooning. Were women overstepping their bounds? The Chief of Police of Paris evidently thought so: in 1795 he decided that there were to be no more Parisian women in balloons. Women could not possibly stand up to the strain of riding in balloons, he felt, and for their own sakes they should be protected from the temptation to fly.

Nevertheless, it was in France that the next ballooning heroine was found. Sophie Blanchard's fascination with

6

balloons began when her husband, Jean-Pierre Blanchard, became interested in them himself. Blanchard's specific interest was in developing a balloon that could be controlled and propelled by its rider. He was totally unsuccessful, as it turned out. The balloon he developed was equipped with meaningless "airscrews," and also with a pair of "air oars" with which to row through the skies. However, Blanchard's failure at balloon design was soon forgotten, for in January, 1785, he and a partner became the first people ever to cross the English Channel in a balloon. Later Blanchard, now internationally famous, would make the first balloon voyage in the United States, in the presence of George Washington.

Towards the end of his life, Jean-Pierre Blanchard made some bad investments which brought him financial ruin. On his deathbed he told Sophie that she could choose between drowning herself or hanging herself after he was dead, but that he saw no other choice nor any hope for her.

Instead, Madame Blanchard—who had also become a skilled balloonist herself during her husband's life—decided to make his career her own. Her many exhibition voyages, during which she parachuted fireworks out of her balloon, brought her great fame. More important, they gave her a reputation for bravery and coolheadedness. In 1804, Napoleon himself named Madame Blanchard to the post of chief of France's new Aeronautic Corps. (The previous chief had been fired when a balloon he'd designed to celebrate Napoleon's coronation had landed ignominiously on the corner of Nero's tomb.)

Madame Blanchard brought great zest to her new job. When the new King of Rome was born in 1811, she dropped birth announcements from her balloon. She was in charge of many other public festivals as well. In all, she made more than fifty balloon voyages. Many of them were terribly dangerous, but Madame Blanchard was as much of a showman as her husband had been—and the more dangerous the act, the larger the crowd.

Her final flight came on July 6, 1819, at the Tivoli Gardens in Paris. For this flight, Madame Blanchard had had the idea of setting off fireworks, not by parachuting them but by fastening them to the balloon itself, lighting them with a long torch, and then cutting them loose. As an onlooker in the crowd later wrote: "The balloon rose splendidly, to the sound of music and the shoutings of the people. A rain of gold and thousands of stars fell from the car as it ascended. A moment of calm, and then an un-expected light appeared . . . increased, then disappeared suddenly; then appeared again, in the form of an immense jet of blazing gas. The spectators cried 'Brava! Viva Madame Blanchard!' thinking she was giving them an unexpected treat."[1]

What had actually happened was that the torch's fire had ignited the hydrogen balloon. Thousands of spectators, un-aware of any danger, watched as the flaming balloon flew down through the air, smashing against a rooftop. Madame Blanchard plunged to the street and died there of a broken neck. The crowds were still cheering as she died.

It took a comparatively long time for American women to begin ballooning. Not until 1825 did a Mrs. Johnson make an ascent in the United States. Little is known about this first ascent. The New York *Evening Post* reported that Mrs. Johnson was about thirty-five years old and was dressed in a white satin dress and a short red jacket. "She gave the word to let go, bade her friends farewell, waved her flags, and rose with great rapidity, amidst the shouts of the surrounding multitude."[2] Her balloon took Madame John-

---

[1] Dwiggins, Don, *Riders of the Wind: The Story of Ballooning,* p. 34.
[2] Ibid., p. 51.

son across the East River and Brooklyn and dumped her unceremoniously in a marsh, which must have done some damage to the white satin dress.

In 1855 another American woman, Lucretia Bradley from Pennsylvania, made on ascent which was considerably more dramatic. Eager to make a name for herself, Bradley bought a secondhand balloon from the famous American balloonist John Wise. She quickly put together a hydrogen-producing system for the old balloon and made the ascent in January.

The balloon rose easily, and Lucretia Bradley was amazed by the wonderful view it gave her of the countryside—so amazed that she forgot to check the balloon's progress until she was two miles up in the air. When she did glance up, Bradley realized suddenly that the balloon had swollen almost as far as it could go without exploding. She pulled the safety valve to release some of the balloon's hydrogen, pulled it again, and watched in terror as the balloon continued to swell and to float relentlessly higher. Finally she pulled the safety valve as hard as she could, and at last the balloon slowed down as the gas began to escape. It could not escape fast enough. Two or three seconds later the balloon exploded and began to drop at the rate of 100 feet a second.

By miraculous luck, Lucretia Bradley did not die in the accident. In fact she landed unhurt, though breathless: the falling balloon had formed a parachute. Bradley walked away from the scene and later wrote a letter to John Wise. She wondered if he could sell her another balloon.

Ballooning remained the only way to travel by air for more than a century after the first balloon had flown. The fact that balloons were impossible to steer, controllable only by capricious winds, meant that no balloon flight could ever be completely safe. This, of course, made the flights more interesting to the public. Ballooning in the nineteenth century was mostly a spectator sport, popular with crowds

at circuses and fairs. Balloonists were looked on as enter-
tainers, and never more so than in 1874, when the first
aerial wedding in the world took place.

Miss Mary Elizabeth Walsh was the bride. The balloon,
appropriately, was the "P. T. Barnum." The bride—an
equestrienne with Barnum's Roman Hippodrome—wore "a
delicate pearl-colored silk, with bias folds and heavy trim-
mings of fringe and puffing in the back." The bridesmaid
wore black. The balloon was specially decorated with bou-
quets, flags, and ribbons to make it look as romantic as
possible. The wedding party rose more than a mile before
the ceremony was over. The minister announced weightily
after the ceremony that, since marriage is "an institution
above those of the world, merely, it is, then, most fitting
that its solemnization should be celebrated far above the
earth."[3] The whole event is worth remembering as one of
the most frivolous in the history of ballooning.

To the modern mind, balloons may seem like fickle,
ridiculously fragile contraptions useful only for providing
adventure. But that is because the twentieth century takes
flight for granted. It is hard to realize what ballooning
meant in its early days. Balloons embodied one of history's
oldest dreams. Those flimsy contraptions proved that ob-
jects heavier than air could become lighter than air, that
people were not tied to the sober earth. Whether they
flew for adventure, for the public, or for romance, the
women who traveled in balloons should be remembered
for more than their courage. Like the balloon itself, they
started a new age.

THE HISTORY OF PARACHUTING closely followed that of
ballooning. This was quite natural. Once people knew that

[3] Lt. Colonel Glines, ed., *Lighter-than-Air Flight*, pp. 76–77.

they could be put into the air, they sensibly wanted to be sure they could return in one piece, and the first parachute was created less than two years after the first balloon.

Jean-Pierre Blanchard was the creator. His unwilling test subject was a dog lowered in a basket attached to a parachute. This took place on a summer balloon voyage in 1785, and although the dog was probably only concerned with getting down safely (which he did), the incident intrigued people. With parachutes aboard, balloonists could not be overpowered by even the most capricious winds. Ballooning —and flight—now seemed easier to control.

After Blanchard came Andre-Jacques Garnerin, a Frenchman who made the first manned parachute jump in history. Garnerin became famous as both a parachutist and a daredevil balloonist. (On one night ascent he went so high he saw meteors flashing past him, or at least he said he had.) Garnerin also allowed his niece Elisa to come with him on some of his ascents. Elisa was one of the first women to parachute, and she was the most famous of her time.

Though parachuting may have seemed safe in comparison with ballooning, it is hard to believe how flimsy the first parachutes were. To make a descent, Elisa Garnerin had to climb into a tiny gondola attached to the larger balloon gondola. Next she pulled out a knife and cut the ropes tying the two gondolas. Even if something went wrong at this point, the parachutist could not be helped by the balloonist. Once the ropes had been cut, the parachute gondola instantly began to drop, and as the parachute fell, the balloon, now so much lighter, suddenly jumped up. If everything went smoothly, the tumbling parachute gondola then shook the parachute open, and Elisa bobbed slowly down to earth.

Despite all the dangers, Elisa Garnerin was a confident parachutist. She was confident enough, in fact, to make bets about where she'd land, and to win them.

Elisa Garnerin was rivaled by a countrywoman of hers,

Madame Pointevin. Madame Pointevin was the wife of a stunt balloonist herself, and it is possible that she parachuted less for the sheer joy of it than for the cash it brought her. At any rate, she did make good money, which seemed scandalous to many people at the time. The English were especially displeased. For a woman to jump in the first place, and on top of that to do it for pay! It was decided that Madame Pointevin would not be permitted to disgrace herself in England.

But she was a resourceful woman. She told the authorities about a famous trip she had taken in France. She had made an ascent in a balloon tied around the stomach of a horse and had come down perfectly unharmed. If she could do that, Madame Pointevin urged, she could certainly manage a simple parachute jump without causing trouble. Moreover, she wanted to try an ascent while riding a bull as well. The English gave in at last. There was to be no aerial bull-riding, but Madame Pointevin could make parachute jumps herself if she wanted to.

Margaret Graham was an Englishwoman who made several balloon ascents in the middle of the nineteenth century. She too caused plenty of scandalized gossip, for she often took her three daughters up with her. On one bizarre occasion, Mrs. Graham made use of a parachute without having planned to. She was in a balloon at the time, and to her horror it suddenly collapsed, spilling her out of the gondola. Fortunately, this was the era of ankle-length skirts and billowing petticoats. As Mrs. Graham plummeted down feet first, her skirts and petticoats filled with air to form a crude parachute. She landed safely, with only her dignity bruised.

The German Käthe Paulus was one of the most famous women to parachute in the late nineteenth century. Like Madame Pointevin, Paulus made stunt jumps, wearing boots and bloomers instead of a long skirt. What made her act different was that she jumped with her fiancé. Crowds loved this. Somehow it made the jumps seem far more dangerous; and it made the couple's tragedy, when it came, seem even more terrible. On one routine jump, Käthe Paulus's parachute opened normally. As she descended slowly through the sky, she looked around for her fiancé—and saw him hurtling through the air much too fast. His parachute had never opened, and Käthe had to watch, helpless, as he crashed far below her. Paulus continued to jump after the accident, although she had been urged to stop for good.

Parachuters rode in baskets attached to the actual parachutes almost until the twentieth century. Ada MacDonald was one of the first people to jump with the parachute attached to herself. She ascended sitting calmly in a chair tied to a balloon. The parachute was also tied to the balloon. As soon as she was high enough, MacDonald slid off the chair and, with the chute sewn to a heavy strap around her waist, she descended. That is, if everything went smoothly. On one demonstration flight in Wales in 1890, MacDonald sat down in the ballooning-chair and signaled the men in charge of the exhibition to let the balloon loose.

Without meaning to, they let go of both the balloon's ropes and the ropes that tied the chair to the balloon. The balloon jumped straight into the air with Ada attached. Now the parachute should have broken away from the balloon, but the two remained firmly tied together.

No one will ever be sure why this happened, but it is very possible that someone had tied them together on purpose. In any case, the ropes were tied and knotted. MacDonald worked frantically to untangle them; in doing so she ripped the parachute loose from the balloon and began to fall. The crowds watching panicked, screamed, and scrambled to get out of her way. She was falling faster and faster toward them, and it looked as though she had no chance. Just as it seemed certain that MacDonald would be killed, her parachute whipped open with a terrible jolt. If someone had been trying to sabotage her, the try had failed. MacDonald had survived, but it took her a long time to recover from the shock of the experience.

One of the most famous American parachute jumpers was Tiny Broadwick. She began to jump in 1908 at the age of fifteen—a fifteen-year-old mother and widow, unable to read or write anything except her own name. She became the foremost parachutist in the nation.

Tiny Broadwick was christened Georgia Ann Thompson. Her nickname came from the fact that she really was tiny, only eighty pounds and less than five feet tall as an adult. Born in the South into a poor family, Tiny began working very young. Her first job was picking tobacco worms off the leaves in tobacco fields; next she married and had a baby, but when her husband died she had to return to work, this time in the cotton fields. One day, on her day off, Tiny went to the carnival in Raleigh, North Carolina. There she saw her first balloon and her first balloonist, Charles Broadwick.

Broadwick's balloon and parachute act entranced Tiny. Right away she decided that she would make a good parachutist: "Something told me that was what I wanted to do. I just stayed there until he was done, and I told him I wanted to join up with him. He said I looked so little. I said, 'Well, my daddy passed away and my mama will take care of my little girl for me' . . . I told him I was a tomboy, Georgia Tomboy, always playing with the boys, climbing trees and falling out of trees. To him, I wasn't a girl, I was a tomboy, so he taught me all I knew, and he had confidence that I would make good."[4]

And she did make good, jumping for the first time after only a week of instruction on the ground. The carnival

[4] Vecsey, George, and Dade, George C., *Getting off the Ground*, p. 32.

packed up and left Raleigh that day. To her disappointment, Tiny never got a chance to see whether the newspapers had mentioned her first jump. There were other papers, though. Billed as "The Doll Girl," Tiny was made to dress like a doll in white dresses and white ruffled pantaloons; posters called her "A Tiny Chance-Taking Slip of a Girl," and she was photographed holding a little parasol, as if that were all she needed to bring her safely through the air.

Tiny suffered through many accidents, falling into trees and ponds and once through the window of a train. She tried never to think about the dangers facing her, saying years later, "I had enough close calls to learn how to handle the chute. If I had trouble, I would work it out myself . . . I didn't talk about danger in those days. If you talked about it, danger would confront you all the time. I thought about the excitement as the people gathered around to watch. The applause was grand."[5]

THE MAIN PROBLEM with hot-air balloons is that they are not dirigible, that is, not steerable. Ballooning has certainly always been thrilling, but early balloonists who landed unceremoniously in cow fields miles from where they had hoped to come down, or who floated up so high that their drinking water froze, must often have wished for more control over their flight.

People began to develop dirigibles as soon as the first balloonists had flown successfully. The main distinction of some of the earliest attempts is that they were overpoweringly silly. In 1850, for example, Monsieur Pointevin enlisted the help of three women in a new balloon he'd designed. The three were specially equipped with wings fas-

[5] Ibid., p. 34.

tened to their shoulders and arms. They went up in the balloon with Pointevin on October 13, and as soon as the balloon was high enough, Pointevin carefully lowered the women one by one on ropes into the air. At this point they were supposed to fly the balloon along wherever Pointevin directed. Of course this didn't work. The women flapped their arms and kicked while the balloon floated serenely on, absolutely unaffected.

In 1898 the first successful dirigible was invented by the Brazilian Alberto Santos-Dumont. It used an efficient gasoline motor rather than flapping women; it could change direction and fly straight into the wind. Now at last people could really choose where to fly.

Santos-Dumont must have been a more tolerant man than many in his day, because in 1903 he decided to allow a woman to fly his dirigible herself. She was a dashing Cuban, Aïda de Acosta, who was living in Paris at the time. Santos-Dumont showed her the basics of flying and soon told her she was ready to fly solo. On July 29, five months before the Wright Brothers made their first flight, Aïda de Acosta made hers.

She did not tell the public what she was going to do; she just took off and flew the dirigible without any problems. It came to rest near a polo field and interrupted the game in progress. Everyone within sight rushed over to get a glimpse of Santos-Dumont (they believed he was the pilot) and found instead a self-assured woman in a black-and-white dress and a sweeping hat adorned with pink roses. "It is very nice," de Acosta politely told Santos-Dumont as he rushed over to see how she had done.

Aïda de Acosta became famous in Paris, but Santos-Dumont was greatly criticized. How could he have been so cruel as to place a woman in such a dangerous position? people asked. They were not convinced when they were assured that de Acosta was perfectly all right. Nor were her parents at all happy. They hated the thought of their daughter's receiving vulgar publicity, and they made Santos-Dumont promise never to reveal his pupil's name. The historical voyage sank out of sight.

Gertrude Bacon became the first British woman to ride in a dirigible in 1904, flying with Stanley Spencer, who had designed the airship. In contrast with Aïda de Acosta's flight, Gertrude Bacon's was well-publicized. More than seventy thousand people were at hand to watch it, and they were furious when Spencer announced that the flight would have to be put off because of stormy weather. In fact, the spectators insisted that the flight must take place, and it looked for a while as though they might riot. Finally the storm abated a little, and Spencer and Bacon took off.

The two voyagers realized immediately that the space allowed for the take-off was too small. A stand of tall trees was on one side of the area; a huge exhibition tent was on the other side; and the tremendous crowd presented a further danger. The dirigible rose and was suddenly caught in a burst of wind. It headed directly toward the tent, and people began to panic.

But Stanley Spencer managed to throw one of the sand-bags out of the airship before the propeller hit the tent. The ship rose quickly—trimming some flags off the tent as it did so—and grazing the trees slightly, veered off into the air.

The rest of the voyage was uneventful, but Gertrude Bacon never forgot it. In her 1928 autobiography *Memories of Land and Sky* she wrote glowingly, "I thought the voyage the most rapturous I had yet known, combining all the rare beauty of balloon travel, the matchless panorama, the space, the freedom, with the charm of motion, the thrilling sense of life. I felt the breeze on my cheek, the vibration of the throbbing engine beneath my feet; I saw the revolving blades, I smelt—yes, certainly that added to it—the burnt petrol that in those days invariably accompanied motor travel, and I felt that life itself had little more to offer."[6]

In 1906 the first American woman rode in a dirigible. Mary Miller took off with Leo Stevens, a pilot. As the airship rose, Mrs. Miller tossed the rope anchoring it out onto the ground. Without her knowing it, the rope snared the engine crank and knocked it off. The dirigible rose; it drifted for a few minutes; and then the engine suddenly stopped. Without the crank, there was no way to start the engine, and the two passengers were helpless. The dirigible hit the side of a house and came to rest unharmed—just a few feet short of a ravine. If the airship had not collided with the house, it would certainly have crashed into the ravine. Luck had made a small crash prevent a bigger one.

Dirigible flight never had time to assume vast importance in the history of aviation. Mrs. Miller flew in 1906. In 1908 the first woman in history flew in an airplane, and that is where the most important accomplishments of women aviators began.

[6] Bacon, Gertrude, *Memories of Land and Sky*, p. 137.

The First Women in Airplanes

IT WOULD BE an understatement to say that learning to fly was a difficult task for a woman in the early days of aviation. To begin with, she would have had a hard time finding a teacher: not many men flew at the beginning of this century, and most of those who did were unwilling to teach women. So she would have had to fight in order to learn to get into the air, and then she would have had to fight in order to stay there. Most nations at this time did not want women to vote, much less waste their time on careers in aviation.

The prevailing attitude toward women and flight was well expressed by an early pilot, Jeanie McPherson, who was a woman herself: "Flying is not an easy thing for most women. They seem to lack the sense of balance necessary to become a good pilot. That undefinable sense of 'feel' which airmen and airwomen must possess is absent. Of direction they know little. Tightly laced feminine apparel and high heels may have had something to do with this . . . Women are not calm in flying. The moment a machine begins to behave badly, the woman becomes panicky, works the controls frantically, and finds herself in a perilous position."[1]

It was no more helpful that one man, explaining why women *should* fly, stated that women had been taught to sneeze so daintily that at least they would be placed in less danger than men if they sneezed at the controls.

The first women to fly ignored such pronouncements. It is hard to know which of the earliest records to believe, but in 1908 two different women are credited with being the first to fly as passengers in airplanes. One was a Mademoiselle P. van Pottelsberghe, who flew with Henry Farman at Ghent, Belgium. The other was Madame Thérèse Peltier, who flew with Léon de la Grange in France. (De la Grange was himself the third Frenchman ever to fly.) In 1908, Mrs. Hart O. Berg also flew in France as the first passenger of the Wright Brothers.

The French also claim Thérèse Peltier as the first woman actually to fly an airplane. She began flying lessons with de la Grange in 1908, shortly after her first voyage. Peltier flew for more than a year, but she quit completely in 1910 after de la Grange had been killed in a plane crash.

The first British woman to fly was probably Mrs. F. S. Cody, who flew with her husband. Mrs. Cody prepared for the flight by tying her skirts securely around her knees and tying her hat to her head with a long scarf. She might be daring enough to fly, but she wouldn't appear in public without a hat.

[1] Roseberry, C. R., *The Challenging Skies*, p. 421.

Gertrude Bacon, who had been the first British woman to fly in a dirigible, went in 1909 to the first international aviation meet in France and realized immediately that she wanted to learn to fly. She announced this intention, but for the most part the pilots at the meet paid no attention, thinking Bacon just wanted to show off.

Only one man, Roger Sommer, decided to give her a chance. On the date they had set for Gertrude Bacon's first flight, she arrived at the hangar in time to see Sommer suffer an accident while landing. But that didn't bother either of them; they just agreed to meet again once the plane had been fixed.

Gertrude Bacon loved her first flight, even though it was hard for her to see much: Sommers's scarf flapped in her face constantly. They flew for an hour, and the motor was so loud that Bacon was deaf for several minutes after landing. Such little inconveniences did not bother her, however. "The ground was very rough and hard," she wrote later, "and as we tore along, at an increasing pace that was greater than any motor I had yet been in, I expected to be jerked and jolted. But the motion was wonderfully smooth —smoother yet—and then! Suddenly there had come into it a new, indescribable quality—a lift—a lightness—a life! . . . You wonderful aerial record-breakers of today and of the years to come, whose exploits I may only marvel at and idly envy, I have experienced something that can never be yours and can never be taken away from me—the rapture, the glory and the glamour of the very beginning."[2]

The airplanes known to Gertrude Bacon and other air travelers of the day were almost as different from modern planes as bicycles from cars. Frail and spare, they were built mainly of wood, wire, and fabric. Two principle types of plane were flown in the first part of the twentieth century.

[2] Bacon, Gertrude, *Memories of Land and Sky*, p. 162.

Biplanes had two sets of wings, the top set connected to the bottom by an airy system of wires and struts (supporting bars); their pilots sat upright on the lower wing surface between the struts. Monoplanes had one set of wings, like modern airplanes; their pilots sat hunched in small open cockpits. Neither airplane type had a windshield, so goggles were a necessity. Neither type offered any protection from the weather. Such luxurious details as closed cabins would not be added to airplane construction for some time.

The first woman to become a licensed pilot was the French Baroness de la Roche, who obtained a license in 1910. (She had flown solo two years earlier.) Her teacher was Monsieur Voisin, one of the first professional airplane builders. He was an excellent teacher and she was an excellent pupil, and it was the fault of neither of them that the Baroness had an accident shortly after she had been given her license. Her plane was caught in the "backwash" of another plane flying in front of her; in the crash the Baroness broke both her legs. She was so unnerved that she vowed never to fly again. "The rapture, the glory and the glamour" weren't worth it.

The Baroness was inspired to start again, oddly enough, when Voisin died two years later in a car crash. She realized that cars were as dangerous as airplanes—or that flying a plane was as safe as driving a car—and she began to fly once more.

Hélène Dutrieu was the second French woman to receive her pilot's license, and she was even more famous in her day than the Baroness. She racked up award after award in a remarkably short time, beginning in 1909 when she flew for over an hour. Next she set a distance record by flying over a twenty-eight-mile course; she set an altitude record by flying one thousand three hundred feet above the city of Bruges; she was the first woman to fly in a 1910 aviation meet, even though no prizes were awarded to women; and in 1911 she made a 137-mile nonstop flight in France. Dutrieu also made an unofficial "first" for women aviators by wearing pants in public when she flew. It was certainly an easier way to pilot a plane than in a long skirt.

The first American woman to make a solo flight was Blanche Stuart Scott, nicknamed Betty, from Rochester, New York. Betty Scott was not encouraged to fly. In fact, her friends were openly discouraging, warning her that she would become dizzy high up in the air and that there would be no steadying male nearby to help her down to the ground. "I do not intend to get myself killed by trying foolish stunts," she answered, and she was so sure of this that she managed to convince Glenn Curtiss, the celebrated American aviator, airplane designer, and inventor, to teach her to fly.

Blanche Scott's first solo flight was on September 2, 1910, but it was unintentional. She was sitting in the plane while Curtiss gave her a lesson. As she began to practice driving the plane back and forth across the field, a sudden gust of wind lifted the plane into the air. Because the engine was on at the same time, and because airplanes in those days were lighter than they are now, the plane was soon forty feet in the air. That is not very high, but it can seem a terrible distance to a novice pilot. Fortunately Scott kept her head, perhaps remembering her staunch answer to her friends' misgivings. She landed the plane smoothly. Glenn Curtiss had been teaching her for only three days, but he decided she was clearly ready to begin flying. He never did become convinced that women should fly at all, although he continued to teach Scott; after that he never had another woman pupil.

Another American woman named Bessica Raiche was learning to fly at the same time as Blanche Scott. Bessica had left Beloit, Wisconsin to study music in Paris and while there had become fascinated by stories about the French aviators. While in Paris she also met and married Frank Raiche, an airplane designer. The newlyweds returned to America, and Bessica began to study aviation instead of music. She was an exceptionally fast learner. Two weeks after Blanche Scott first soloed, so did Bessica Raiche. She could only get the plane a few feet off the ground, but in another two weeks she managed to fly it at a more respectable height.

Because Blanche Scott's solo flight had been made possible by a trick of the wind, the Aeronautical Society decided that the first solo honors should go to Bessica Raiche. It must have been a little hard on Blanche Scott that on October 13, 1910, a formal dinner was given for Raiche by the Society. She was awarded a gold medal studded with diamonds; on it was inscribed, "First Woman Aviator of America, Bessica Raiche."

Raiche then began to devote her time to helping her husband design and construct airplanes. The Raiches were especially interested in making planes lighter, and they were the first to use piano wire rather than iron stove wire in airplane construction. In addition they introduced bamboo and Chinese silk—lighter than wood and canvas—into the American aviation industry.

Eventually Bessica Raiche was forced to give up flying because it was a strain on her health. By the time she had recovered her health, she had lost her interest in flying. Instead she began to study medicine and became a practising physician. In medicine she finally found a lasting career, and she never returned to aviation.

Neither Bessica Raiche nor Blanche Scott had a flying license. Licenses had not been necessary when they were taught to fly. Although the Aero Club soon forbade new pilots to fly without licenses, it did not require existing pilots to obtain them. Blanche Scott thought the requirements for being licensed were too strict to make it worth trying for, so she went on flying without one.

The first American to receive a pilot's license was the strikingly beautiful Harriet Quimby. Quimby was the dramatic editor of *Leslie's Weekly* when she began learning to fly in April, 1911. Unlike some other aviators, Quimby was not boastful about flying. In fact she was determined to keep her lessons on Long Island completely secret. They were secret for a little while, secret in a dashing and romantic way. Quimby worked with her flying instructor just after dawn; she always wore an aviator's suit and shrouded her face in a deep hood. Her cover was ruined when she was forced to make a crash landing after only a few weeks of training. The accident—from which she escaped safely— made all the front pages, and Quimby was discovered. She continued to study, though, until she had received her license in August.

By December, Harriet Quimby had learned enough about flying to be a stunt pilot at the inauguration of the president of Mexico. But she wanted more than that. She began to plan, again in complete secrecy, to be the first woman to cross the English Channel in an airplane. There was a reason for the secrecy this time: Harriet Quimby did not want any other woman to find out her plans and try to beat her across the Channel. She arranged to meet Louis Bleriot for lessons in flying one of the planes he'd designed, and she arranged as well to have a plane like the one she had been flying in America shipped to Dover, where she was planning to take off. Unfortunately the plane was spotted on its way to her. People became suspicious. If the plane was being shipped secretly, someone must have big plans for it.

Eleanor Trehawk Davis, a British woman, evidently suspected what those plans might be. On April 2, 1912, Davis flew across the Channel as a passenger, something that no woman had ever done; the record became hers, and some of the fun went out of Harriet Quimby's preparations.

Now it was up to Quimby to become the first woman actually to pilot across the Channel. She worked furiously for two more weeks before deciding that she was ready. But the weather had never been good enough for her to test the Bleriot airplane she planned to fly in. This was another blow, and to add to it, Quimby was told that if she flew even five miles off course, she would end up in the North Sea. No one thought she should attempt the trip. One pilot (the one who had flown Eleanor Davis across the Channel) told Quimby frankly that she had no chance at all. Wouldn't she like to let him wear her clothes, he suggested, so that he could make the flight in her place? Then he would let her have all the credit.

But Harriet Quimby was not afraid. "My heart was not in my mouth," she said firmly. "I felt impatient to realize this project on which I was determined, despite the protests of my best friends. For the first time I was to fly a Bleriot monoplane. For the first time I was to fly by compass. For the first time I was to fly across water. For the first time I was to fly on the other side of the Atlantic. My anxiety was to get off quickly."[3]

Her clothes for the flight had been chosen almost as carefully as her airplane. She wore her customary purple satin suit, to start with, and under it she wore two pairs of silk long underwear. Then on top came a wool coat and a raincoat, and finally a huge sealskin stole. Just as she was ready to go, her friends forced her to take a hot water bottle as well, which her advisor "insisted on tying to my waist like an enormous locket." Thus laden down, she climbed into the plane and took off.

In half a minute Quimby was fifteen hundred feet above

[3] Planck, Charles E., *Women with Wings*, p. 20.

the choppy Channel waves. Briefly she noticed the crowd of reporters far below her, and then the fog came up and she saw nothing but fog. She was forced to use her compass for guidance, the compass she had never used before. Quimby kept her composure, and in twenty minutes the fog lifted. There below her Quimby could see land and she was filled with disappointment; she must have flown in circles, she thought disgustedly, and have stayed above England the whole time. How could she face the people down on the ground?

But Harriet Quimby was not above England. She was above Calais, she had succeeded, and she was the first woman in the world to have flown herself across the Channel. When she landed on the beach beside a fishing village, the townspeople were so delighted to see her that they brought coffee right down to her plane.

As Quimby wrote later, the detail that had been so annoying at the beginning of the trip—that hot water bottle —had been helpful after all. "I did not suffer from cold while crossing for the excitement stimulated my warmth, but I noticed when I landed that the hot water bag was cold as ice. It surely saved me something."[4]

Harriet Quimby was famous not only for her flying but also for her sense of style. Her purple satin flying costume attracted a lot of attention, and after crossing the Channel she returned to the United States with an all-white Bleriot airplane as well. But she was not to savor her fame very long. On July 1, 1912, less than three months after crossing the Channel, Quimby took a passenger, William Willard, for a short flight near Boston. She was trying to break a speed record, and perhaps it was the plane's tremendous speed that caused the accident.

As Quimby and Willard flew above Dorchester Bay their

[4] Ibid., p. 20.

plane dove downward. Neither of the passengers was wearing a seat belt, and Willard was thrown out of his seat. A second later Harriet fell out too. Both landed in four feet of water and died instantly.

Blanche Scott, flying over Boston, had seen Harriet Quimby die, and she was so horrified that she landed her plane and collapsed.

No one knew what had caused the crash. Perhaps Willard had moved suddenly and caused the plane to lurch; perhaps Quimby had fainted; perhaps a sudden gust of wind had jarred the plane. In any case one of the world's most popular pilots was dead.

The second American woman to obtain a pilot's license was a devoted friend of Harriet Quimby's named Mathilde Moisant. It was at the Moisant School of Aviation, founded by her brother John, that Mathilde learned to fly. (She received her license after only thirty-two minutes' worth of lessons.) Mathilde proved as talented a pilot as Blanche Scott and Harriet Quimby. Only one month after being licensed in 1911, and while making her public flying debut, she set a woman's altitude record by flying twelve thousand feet into the air.

Before the year was out, Mathilde had flown three thousand feet high and had set a new world record. She was the first woman to fly from Paris to London (her brother had been the first man to do the same), and she also flew as an exhibition pilot in Mexico. Despite her reputation as a careful pilot, there were many people who thought she shouldn't be allowed into the air at all. One of these was the Sheriff of Nassau County, who decided he could stop her by arresting her for flying on a Sunday. Moisant simply jumped into her plane, flew away to another airfield, and then zipped off in a car before the Sheriff could get her. The court decided that she was in the right anyway; flying on Sunday was no more immoral than driving a car.

The Moisant family also hoped that Mathilde would stop flying, and their arguments had more weight with her. After John Moisant was killed in an exhibition flight, the Moisant parents begged Mathilde to stop for their sakes.

At last she agreed, but she decided to give herself one more flight before leaving aviation for good. She announced that her last flight would take place on April 14, 1912, at Wichita Falls, Texas. Pictures taken of her around this time showed a tiny, trim woman beaming under a tweed helmet; Moisant always wore a thick, practical tweed uniform—complete with knickers—leather gloves, and high lace-up boots when she flew.

It may have been the uniform that saved Mathilde Moisant's life on April 14. Her flight had gone splendidly, and she was coming in for her landing when a mob of spectators broke through the police barricades and swarmed onto the landing field. If she continued to land gradually, Moisant would not have been able to help mowing into them.

Planes of that era were landed after their motors had been turned off, and it was very dangerous indeed to restart the engines before they had had a chance to cool off. But Mathilde Moisant had no choice. She switched the engine back on and aimed the plane straight down. It hit the ground with a bone-shaking jolt, bouncing sharply back into the air. Mathilde then gunned the engine and managed finally to fly past the crowds. But the effort set the engine on fire. The airplane crashed in a cloud of flame.

Now the spectators who had mobbed the field rushed to save their heroine. They pulled her out of the plane a few seconds later, her clothes completely ablaze. The fire had not been able to burn through the thick tweed of her costume, and the flames were beaten out before they could reach her body.

Mathilde Moisant calmly asked the spectators to inform her family that she was safe. Then she left the field. She kept her word to her parents and never flew as a pilot again, but no flyer could have had a more dramatic farewell scene.

Stunt Pilots & More

N O ONE DENIES that women aviators made great advances at the beginning of the twentieth century. No one denied it then, either, but it was a sad fact that even the most talented women were usually unable to take up serious careers in aviation. Though many of them longed for "regular jobs" as aviators, for the most part they could only earn money by exhibition flying. Women pilots were a novelty, and the public was willing to pay to see them. But to let them teach flying on a steady basis, or work for the government as aviators, or even make deliveries in their planes, was unimaginable. Women's flying was not taken seriously enough to allow them that.

That's why so many women aviators at around the time of World War I made careers for themselves in exhibition flying, performing for crowds. And in the process, some of them managed to work their way into serious flying careers. One of the first women to do this was Katherine Stinson.

Aviation history has often been made by families. The Montgolfier brothers created the first usable balloon; the Wright Brothers invented the first practical airplane; Gertrude Bacon first travelled in balloons with her father; Mathilde Moisant learned to fly at her brother's aviation school. Katherine Stinson was the oldest of four children, all of whom flew, and she was the fourth licensed woman pilot in the United States.

Katherine Stinson was born in Alabama in 1893. Like Bessica Raiche, Stinson originally planned a career in music for herself. She had been interested in aviation since childhood, and she thought that perhaps she could earn enough money as an exhibition aviator to pay for piano lessons. In July, 1912, she received her license, and she began exhibition flights a year later. By then she had decided that aviation, not music, was the path for her.

Katherine Stinson was determined that her exhibits would be as good as any man's; though the public would probably have paid just to see her fly around in circles, this would not have satisfied her at all. If men had looped the loop in airplanes, then so could she. In fact, she may have been the first woman in the world to loop the loop, but it was far from being her only stunt. She also invented a stunt she named the "Dippy Twist Loop," in which she both looped the plane and made it flip wing-over-wing.

Stinson was also the first woman to practice skywriting, and skywriting at night at that. She accomplished this by flying with magnesium flares attached to her plane. They burned so clearly in the dark that they could even be photographed. In one of her best-remembered exhibition events, Stinson performed aerial spiral twists and a loop on a damp, foggy night above the Sheepshead Speedway on Long Island. When she landed, she confessed, "The air bothered me a lot. After I left Coney Island I didn't know just where I was, for my fireworks had blinded me and I was afraid I wouldn't find a good place to land. I knew the Speedway was plenty long enough, but I wasn't sure just where it was."[1]

In September, 1913, Katherine Stinson became the first woman ever to carry airmail. She was performing at the time at a fair in Helena, Montana. The postmaster there had requested that an airmail route be set up from the fairgrounds to the Federal Building in the center of the city. When the request was granted, Stinson was chosen to be the pilot. She flew the special route for several days and carried over a thousand pieces of mail.

Katherine Stinson was extremely popular with crowds. Not only could she fly wonderfully, but her flying seemed even more impressive because she looked so young and delicate. She was one of the first women to fly in Canada, where she made a tremendous hit, and she was certainly the first to fly in China and Japan. Airplanes were rare enough in the Orient; the fact that a woman was piloting one seemed spectacularly unusual there.

The President of China, Li Yung Hung, was especially charmed by Stinson. He gave her a check for thousands of dollars after one of her flights and also presented her with a silver cup. In addition to this, Li Yung Hung asked Stinson—whom he called the "Granddaughter of Heaven"—to perform especially for him and his guests at the Imperial Palace in Peking.

It was while Stinson was performing exhibitions in Japan that she heard that the United States had entered World War I. Immediately she volunteered her services as a fighter pilot to the United States Government; just as quickly they notified her that she wasn't wanted. But Stinson had

[1] Adams, Jean, and Kimball, Margaret, *Heroines of the Sky*, p. 33.

33

grown accustomed to such rejections. She flew home anyway to see how she could help out and realized that at least she could be useful by flying for the Red Cross.

So Stinson flew on Red Cross fund-raising drives and on Liberty Bond drives. She managed to collect two million dollars in pledges. At the same time she managed to set several new women's aviation records. In 1918 she flew for 610 miles without stopping, from San Diego to San Francisco, setting a United States record for both men and women.

Katherine Stinson described that long flight as one of the greatest experiences of her life. She set out at sunrise, after a breakfast of one half of a boiled egg. (A waitress had told her that any woman who wanted to fly could fly hungry.) "I left San Diego at 7:31 and headed north through the heavy fog banked over Los Angeles. I thought first of the record I was going to break, but once in a while thoughts of the spring fashions and that sassy waitress in San Diego came . . . Ofter I could see big caterpillar tractors plowing below, and my thoughts went back to the women working in the fields of Japan. Towns, cities, farms, hills, and mountains passed rapidly. The cold head wind blew into my plane; it cut my lips and chilled me, but I never had any fear. The main thing was speed. I carried along my knitting, but I did not have a chance to do much of it."[2]

When she reached San Francisco, Stinson landed between two long rows of soldiers. Their cheers made her cry.

She made the 610-mile record flight for the sake of her career, just to set a record. Most of the rest of her time between 1917 and 1918 was spent on her war work, although Katherine also found the time to help her sister Marjorie teach at the Stinson School, which was owned and operated by the two sisters and their mother. But she was still eager

[2] Ibid., p. 42.

3 4

to help further in the war, and still she was denied permission to fight. Finally Stinson went to France as an ambulance driver. This job put more of a strain on her than anything else had. She caught influenza and had to return to the United States, her health completely broken.

Katherine Stinson settled in New Mexico, where the air suited her best. There she married a former ace flier and retired from aviation. She was only twenty-five when she stopped flying, and it was hard to believe how much she had packed into the six years of her career.

Bessie Coleman was born in the same year as Katherine Stinson, in Atlanta, Texas. Whatever obstacles and loneliness Coleman faced as an aviator were made much harder by the fact that she was black—the first black woman to fly.

As if her race and sex were not challenging enough, Coleman's childhood was one of deep poverty. When she was only seven, her father, whose ancestors had been American Indians, deserted his family in search of Indian Territory in Oklahoma. Bessie had to start picking cotton to help earn money for the family; later she worked as a laundress as well. She also managed to continue her schooling and even to save a little money. This Coleman used to enroll in Langston Industrial College in Oklahoma. But her small funds were only enough to pay for one semester, and after that she moved to Chicago to look for something else to do.

In Chicago she decided to undertake a more practical education and enrolled at a beauty school. Not surprisingly, Coleman found that being a beautician was dull. Instead of working in a barber shop, she decided she would learn how to fly. This, she soon found, was an almost impossible task. What aviation school was willing to teach a black woman to fly? Apparently none—at least not in the United States. Coleman was told, though, that perhaps European aviators would be more welcoming. She began immediately to study French and, soon after the first World War, armed with a new language and all her savings, she set out for Europe. Here she was much more successful. After two European trips and much training with both French and German instructors, Coleman returned to Chicago.

Now that she had an international pilot's license, people were finally interested in seeing Coleman in the air. Thousands of them paid to see her first exhibition in 1922 at the Checkerboard Field in Chicago. It was the beginning of an extremely popular American tour and of a career that seemed assured.

Flight was not without its setbacks, as Bessie Coleman found. For one thing, her family did not enjoy having her fly; for another, she suffered many broken bones and other injuries after one plane crash. But she was undeterred: "If I can create the minimum of my plans and desires, there shall be no regrets." Now that she had proved she could fly, she wanted to do more. Coleman hoped to found a United States flying school for black aviators.

She would never be able to do so. On April 30, 1926, Coleman made her last flight. It was a test flight, performed just before an exhibition for the Negro Welfare League of Jacksonville, Florida, and it was made in a plane whose engine had failed twice only days earlier. Coleman agreed to test the plane with her agent William Wills, but she was not worried about the flight. In fact she was so unconcerned that she decided not to use the seat belt or parachute that might have saved her. The plane, which Wills was flying, flipped over in midair during a nosedive: its controls had jammed. Coleman plunged almost a mile and died instantly. Wills had worn a seatbelt, but he too was killed when the plane smashed to the ground.

Though Bessie Coleman did not live to establish her own flying school, her influence remained after her. Coleman's career was the inspiration behind a group formed by three women in 1975 to help other black and minority women fly. Appropriately, the group named itself the Bessie Coleman Aviators.

Katherine Stinson's greatest competitor—and also one of her close friends—was Ruth Bancroft Law. Law, the fifth American woman to be a licensed pilot, belonged to another family that considered aviation very important. Her brother Rodman had a reputation for being one of the most daring stunt men of the day. This fact may have helped Ruth to find her teacher, Lincoln Beachey, who was a remarkably talented stunt flier himself. Beachey had flown at a record altitude and had flown over Niagara Falls. He was also known for switching his engine off high in the air, just for fun. Beachey was a daredevil, and his pupil Ruth Law became one too.

Ruth Law needed extra courage to begin flying. While waiting for her first airplane ride, Law had seen Harriet Quimby tossed out of her plane and killed. A woman less determined to be a pilot might have given up at that point, but Law knew that all aviators must face similar risks. One month later, on August 1, 1912, she made her first solo flight. Only a few months after that, Law and her airplane had become a familiar sight on the East Coast.

In 1913 Ruth Law married Charles Oliver. There is no doubt that he helped her greatly with her career. Oliver was so proud of her flying that he became her manager, and he turned out to be an ideal pilot's manager: a skilled businessman, an ingenious mechanic, and an enthusiastic publicist. Charles Oliver was a fan of his wife's, and he wanted other people to admire her as much as he did. Because of this he gave wonderful interviews.

But of course Law's own talent was what set off her career. In 1915 she had flown upside down, given an exhibition at night, and looped the loop. In Birmingham, Alabama, she made sixteen loops in a row. Ruth Law may have been the first woman to make a loop; no one knows whether it was she or Katherine Stinson. In any case, the two women often combined forces once they had both become famous for stunt flying, touring together in many exhibitions. Law also made over nine hundred flights in Florida alone, and she earned extra money—fifty dollars a trip—by flying famous people who wanted the status of a trip with her.

In 1916, Ruth Law was chosen to perform in exhibitions at the Iowa State Fair. That may not sound terribly exciting now, but the fair was one of the biggest in the Midwest, and competition to fly there was very great. The Fair commissioners wanted to be sure of a good show, so they told Law that she would receive no pay until she had completely satisfied them and the spectators. Moreover, her contract required Law to perform all her stunts in front of the grandstand, where everyone could see her up close.

It was because she had to fly so near the grandstand that Law almost had a terrible accident. While she was making the top of a loop (which meant that her plane was upside down) her engine died. Law quickly turned the plane right side up and prepared to make an emergency landing. But there was no space to land in front of the grandstand. The only possible place to land was on the green in the center of a race track, and there was a car race going on; if Law landed too close to the edge of the green, she would be powerless to stop the plane from rolling right into the path of the racing cars. She had to brake the airplane the minute she landed. Fortunately she was quick enough to save herself. Her plane rolled straight to the edge of the track and came to rest there.

This was certainly a trick worth paying for! Law's audience went wild. They thought she had created that landing just for them, and the fair commissioners paid her without any trouble.

In addition to stunt flying, Ruth Law was interested in setting records for both distance and altitude. They too brought her great fame. She was honored with a dinner at the Waldorf Hotel which President Wilson and his wife attended, and the Aero Club of America held another formal dinner at the Astor Hotel, where Law was given a check for $2,500. One of the less pleasant benefits of being famous, though, was that advertisers pestered her constantly to get her endorsements for their products. One producer even offered to pay her royally for riding a motorcycle upside down on a Broadway stage. That made her furious. As she said, such requests proved that pilots, especially women pilots, were still thought of as circus people.

It must have been almost a relief when Ruth Law began flying to help the United States in WWI. Like Katherine Stinson, Law offered herself as a fighter: "I could drive a machine with a gun and gunner and go into actual battle. That's what I'd like to do more than anything—get right into the fight."[3] But this offer was refused as Stinson's had been. (In France, Hélène Dutrieu had been accepted for active war service at once.) Instead Law began to fly for the benefit of the Red Cross and Liberty Bond drives.

Still she was not satisfied. In November, 1917, she sent a petition to the War Department requesting that she be enlisted in the Air Force. Once again she was turned down, but at least this time she was allowed to do recruiting work for the Army and Navy. For that work she was given an Army Officer's uniform. She was the first woman ever to wear one.

After the war Ruth Law returned to a private career. She performed stunts, toured the Orient, and delivered the first airmail to Manila. She also founded a flying circus with her husband.

In one of the special acts of the circus, performers used rope ladders to jump from cars onto flying airplanes. In

[3] Ibid., p. 62.

Many of the first women to make aviation history were French. One of these was Madame Thible, an opera singer who became the first woman to travel in a balloon. The historic voyage took place on June 4, 1784, above hundreds of onlookers in Lyon.

This charming portrait shows the famous balloon voyage of Letitia Ann Sage—the first British woman to make a balloon ascent—and her two companions. There is no doubt that the picture is idealized. Vincent Lunardi, shown at right, was actually forced to remain on the ground because the balloon could not lift its three-passenger cargo. The voyage ended in a vegetable field.

At the age of fifteen, Georgia Ann Thompson—or Tiny Broadwick, as she came to be known—left the cotton fields of North Carolina to make parachute jumps with a traveling carnival. She became one of the foremost American parachutists and even tested parachutes for the Navy.

In 1908, Thérèse Peltier of France became one of the first two
women ever to fly in an airplane; her pilot, shown here, was
Leon de la Grange. In the same year Peltier started flying
lessons with de la Grange, becoming the first woman in history
to pilot a plane.

Baroness de la Roche claimed another aviation distinction for
France when she became the world's first licensed pilot in 1910.

In the 1930s Ruth Nichols excelled at altitude flying—on one occasion flying so high that her tongue froze. Nichols was considered one of the decade's most glamorous pilots, partly because of her social background and partly because she photographed so well.

Blanche Stuart Scott poses tensely in a grounded airplane. Scott was the first American woman to make a solo flight, a feat she accomplished when the wind lifted her plane into the air unexpectedly as she was driving it along a field.

The dashing Harriet Quimby was the first woman to pilot across the English Channel (having refused the offer of a male pilot to pose as Quimby for the flight and then give her the credit). Quimby's great beauty and purple satin costume attracted as much public attention as the flight did.

COMPLIMENTS OF
MISS HARRIET QUIMBY

The Stinson sisters, Marjorie (at left) and Katherine (at right) owned and operated a flying school along with their mother. This alone set them apart from most women of the period, but the sisters were also famous pilots in their own right. Katherine Stinson, for example, was the first woman to fly in the Orient and the first woman to deliver airmail, as well as the first woman to practice skywriting.

Mathilde Moisant, the second American woman to obtain her pilot's license, insisted on an extremely sensible flying costume made of thick tweed—including a tweed helmet.

Texas-born Bessie Coleman was the first black woman to fly an airplane. Because she could find no American willing to teach her, Coleman learned to fly in Europe shortly after World War I, returning in triumph to make her first public exhibition in 1922 in Chicago.

1922 a stuntwoman named Maxine Davis tried this act. While Ruth Law drove the speeding car, Davis leaped out to grab the rope ladder. She caught it, but her arms were too weak to pull her up, and Davis plunged to the ground. The fall was fatal.

Ruth Law was horrified at the accident, and the strain following Davis's death was so great that Charles Oliver had a nervous breakdown. He had helped Law devotedly with her career for many years; now he begged her to stop flying. Law had been preparing to make a flight across the Atlantic, but at Oliver's request she gave up flying immediately and moved with him to Beverly Hills. Years later a reporter asked Ruth Law if she would ever like to fly again. Law answered that she had no interest in leaving the ground. Modern pilots had so many rules to follow, she said, that she couldn't imagine how they had any fun.

Phoebe Fairgrave was another woman who began as a stunt aviator. She had said, "I never want to do it again" when she made her first record-breaking parachute jump, but when she realized how much money she could make from such stunts, she decided to try again. Aviation at the time, she said, was "all flying and not an industry," and if the only way she could earn a living in it was by stunt flying, then she would do that.

Like so many other women aviators, Phoebe Fairgrave did not plan to be an aviator originally. In fact, she went to business school to learn stenography. After graduating in 1920, Fairgrave got a job as a stenographer, and before two weeks had passed, she realized that she never wanted anything to do with stenography again. She was seventeen at the time, and she had watched Ruth Law perform; that was enough to convince Fairgrave that she wanted to learn to fly, too.

She paid for three aviation lessons and then came into her fourth lesson with some good news: she had inherited a little money, and she wanted to buy the plane in which she'd been learning. When the lesson was over, the plane had been sold to Fairgrave for $3,500. Phoebe's teacher also saw to it that he was paid for her fourth lesson.

Now she had the plane, but Fairgrave was a little worried about how she would explain the purchase to her family. She decided that they would be happier if they could be sure the airplane would actually be of any use to her. The next morning, she marched over to the Fox Moving Picture Company and managed to sign a contract with them. She would perform aerial stunts for their movies—parachuting and wing-walking—for the price of $3,500, just what she had paid for her plane.

Everything was all set except that Phoebe Fairgrave had never done any of the stunts she'd agreed to do. At her fifth aviation lesson, she told her instructor that she wanted to make a parachute jump. He refused to let her jump with so little flying experience, but another pilot, Vernon Omlie, volunteered to take her up and let her try. Fairgrave made the jump from the end of the airplane's wing; she had wanted to get some practice in wing-walking. The jump went fine, although she landed in a tree and had to be untangled. Her stunt career had begun, and so had her life with Vernon Omlie, whom she married in 1922.

Phoebe Fairgrave's stunts were featured in the Fox motion picture series called "The Perils of Pauline." Along with another stunt man, Glenn Messer, she and Omlie also established themselves as a touring exhibition team, and the exhibitions they gave were spectacular. Fairgrave hung by her knees and teeth from flying airplanes, she walked all over airplanes while they were high in the air, and she even did the Charleston on the top wing of the plane while the pilot looped the loop. Wing-walking would be impossible on modern planes, but it was not as difficult as it looked when Fairgrave performed the stunt. For one thing, the biplane's wings then were held together with bracing wires and struts, so Fairgrave could always grab onto something if she needed to. As she said in an interview, "You just shinny up the strut, grab hold of something on the top wing, throw your knee up there, and climb up." For another thing, the planes in her day flew much more slowly than they do now. Still, wing-walking, even if not terribly difficult, was extremely dangerous.

But it was not nearly as dangerous as another stunt Fairgrave invented: changing from one moving airplane to another in midair. Phoebe, Omlie, and Messer practiced this trick for weeks using only a horse and buggy. Messer would hang from his knees on a trapeze bar in an old barn; Omlie would slowly drive a team of horses underneath him; and Fairgrave, balanced precariously on the seat of the buggy, would grab Messer's hands and be pulled up. The first time they tried the trick with planes, the lower plane almost hit the upper. Its propeller narrowly missed chopping into Fairgrave. To avoid having that happen again, the three stunt pilots decided that in the future Messer would hang on a twenty-foot rope ladder to grab Fairgrave. That way the planes would have no chance to collide.

Phoebe Fairgrave and Vernon Omlie had always been more interested in serious aviation than in stunts, despite their success in exhibitions. Stunts, they felt, were useful enough in getting the public to pay attention to pilots. Now it was time to get people to think of aviation as a serious industry. In 1921, Omlie was able to prove how useful planes could be when he spotted forest fires in his plane. In one flight, he found eighty-seven fires and was able to warn firefighters where the fires were most likely to spread.

Further proof of how helpful airplanes could be came with the terrible Mississippi floods of 1925. Fairgrave and Omlie had been living in Memphis at the time, trying to make the city build an airport. Now they had a real opportunity to be useful. Fairgrave flew with Omlie and two other men for eight days. They brought medicine and food to stranded flood victims, ferried mail from Memphis to Little Rock, and patrolled the levees ceaselessly, checking for new flooding. Several times she and Vernon had to spend nights on the second floors of abandoned, flooded-out houses. When the floods finally receded, the citizens of Memphis had learned that airplanes were not just circus toys, and that pilots were more than show people.

In the Depression years Fairgrave was able to earn an excellent living by racing in air meets, but her husband disapproved: she might be having a good time, he said, but she certainly wasn't doing anything to help aviation. In 1932, she went back to serious airplane work, piloting a woman speaker all over the United States to campaign for Roosevelt. In addition, Fairgrave took a course in public speaking so that she herself could campaign for Roosevelt among aviators.

After Roosevelt was elected, he remembered all that Phoebe Fairgrave had done for him, and she was given a government job. She served as a technical advisor, working as a link between the National Advisory Committee for Aeronautics and the Bureau of Air Commerce. Not only was it a wonderful job, but it also made Phoebe the first woman ever to serve as a government official in aviation. She showed her gratitude in turn by campaigning for Roosevelt again in 1936. But in the same year, Vernon Omlie was killed while flying as a passenger. Fairgrave was so grief-stricken that she gave up aviation entirely.

In 1941, though, she returned to the field in an even more important job for the Civil Aeronautics Administration. Her task was to organize classes to train men for ground work at airports. She had made aviation the most necessary part of her life, and she found that she could not abandon it permanently. Like Katherine Stinson and Ruth Law, Phoebe Fairgrave had helped to make flying respectable, and by doing so she had made herself indispensable to aviation.

The Record-Setters —
Altitude & Endurance

IN 1916 Ruth Law decided to set a new world altitude record in her plane. She had never tried this before, but that fact didn't worry her. She was determined to break the existing women's altitude record, held by Hélène Dutrieu, and the men's world record as well. When she took off at Sheepshead Bay, New York, Law felt completely confident.

As her mother, her brother, and her husband watched, Ruth Law soared into the air. Soon she could not see them any more, and still she continued to fly higher. It was very cold in the plane. As Law climbed higher, she became colder and colder. She knew this meant she was very high indeed, and she was tempted to continue further. But she was alarmed to realize suddenly that her fingers were frozen. She could hardly move them at all, and she could not hold onto the airplane's control stick. If she flew any higher, Law thought to herself, she would lose control completely. She had no choice but to return to earth.

When she landed, Law was unable to get out of the plane. Her husband and brother pulled her out carefully and helped her warm up. As soon as she could talk clearly, Law asked what the barograph read. Her husband told her that she had flown 11,200 feet—twice as high as Hélène Dutrieu had flown to set her record.

Law had broken the women's world altitude record without any question. It would be some time before another woman would beat her. But she had not broken the men's record, and that was what she had really hoped to do. Law was furious. She was so angry, in fact, that she decided she would have to set a cross-country distance record before she could forgive herself. Later in the same year, that is what she did.

It was determination like this that drove women aviators to set records. The pressures against such women were tremendous. They faced danger constantly: setting every new record meant facing a more dangerous flight. But it was this danger which was part of their inspiration. Setting a new record proved to any aviator that she was the best in her field, and it proved that she was as brave as she was skillful.

The French woman Louise Favier broke Ruth Law's altitude record in 1920, flying at 21,325 feet in thirty-five minutes. But not until 1929 were women able to persuade the Fédération Aéronautique Internationale that women who attempted to set records should be judged in a separate category from men. It may seem that this was a step backwards for women, but it was a necessary step at that time. Women had not had as much flying experience as men, and it was harder for women to obtain the excellent planes needed for setting records. Being placed in a separate category gave them the chance to compete against each other and not against male pilots, who had been given better chances to fly from the beginning.

Procedures for official recognition varied (and still do vary) somewhat from country to country, but they were fairly standard. To qualify for an official record, pilots had to apply to a national contest board for official sanction, arrange for a qualified aviator to be approved as their official director for the flight attempt, and make sure that this director report the flight results to the contest board immediately.

The women who attempted to set records in the 1920s, 1930s, and 1940s were usually lent airplanes by plane manufacturers who wanted the publicity a record flight would bring them. Often record-setters received free fuel as well from oil companies. In return the companies used the record-setting events in their advertisements.

Altitude and endurance flights are similar in being extremely grueling. This may be why many women aviators excelled in both areas before World War II; perhaps mastering the strain of one type of record helped them master the other. Though altitude flights are probably more exciting to the public, it might be said that an aviation endurance record is the purest form of record-setting a pilot can achieve. Those who fly for endurance must spend hours confined in a small space with nowhere to go and nothing interesting to see, no new height or speed to strive for, nothing except fighting sleepiness to make the hours more exciting—flying only for the sake of staying in the air as long as possible. It was just as well that the women who set endurance records made names for themselves in other areas of aviation too.

A woman with the charming name of Viola Gentry set

the first important women's endurance record on December 20, 1928. Her only noteworthy aviation accomplishment before that had come when she had flown under the Manhattan and Brooklyn Bridges with a male pilot in 1926. But her endurance record was something meriting more substantial recognition. Gentry stayed up for eight hours, six minutes, and thirty-seven seconds, despite the discouraging words of the man who had told her to stay at home and raise children even as he was helping her into her parachute.

The woman who broke Gentry's record did so only two weeks later. She got the new year off to a good start by staying in the California air for twelve hours and seventeen minutes on January 2, 1929. Her name was Evelyn Trout, but she was more often called Bobby.

The woman who took away Bobby Trout's record—and who did so almost immediately—was Elinor Smith, who was still in her teens. She too had a nickname: "The Flying Flapper of Freeport." And although she hadn't been flying for very long, her early career was as dashing as her nickname.

Elinor Smith had been born in 1910. She first flew as a child when her father bought his own plane and took her for rides. He also encouraged her to learn to fly for herself and made sure she had one of the best teachers—Russell Holderman, who had worked for the Air Mail Service. Smith made her first solo flight when she was only fifteen; this was so unusual that it caught the attention of the famous pilot Charles Lindbergh, who asked to meet her. In 1928 Smith received her transport pilot's license. Once she had it, she celebrated by establishing a new official women's altitude record.

Like Ruth Law, Smith climbed so high that she became chilled to the bone and very dizzy. But she did not stop until the plane's gas gauge read "Empty." Then she began to descend, using the two gallons of gas in the reserve tank. When she landed, Smith had used up every bit of spare gas and had set a new world record of 11,663 feet. (Though Louise Favier and Bertha Horchum had flown higher, their flights had not been officially registered.) It was a remarkable achievement for an eighteen-year-old.

Of course Elinor Smith became famous immediately, and the fame went to her head just as fast. Her next aviation achievement, one month later, was flying under the four East River bridges. For this she won no award; instead a judge ordered her grounded for fifteen days. The Department of Commerce also sent her an official letter censuring her for the stunt. (Included with the letter was a handwritten note from the secretary who had typed it. She wrote that she was proud to see a woman flying, and she wondered if she could have Smith's autograph.)

Eager to regain a more professional reputation, Smith now decided that a good way to do so might be to beat Bobby Trout's endurance record. Smith planned first to set out for a new record only two or three days after Trout's flight, but the weather was so forbidding that she was unable to do so. By January 31 Smith had decided to give up thinking about the endurance record for a while. The weather had not improved yet, and the next day was supposed to be no better. Instead of worrying about flying, she thought, she would go to a dance.

Smith returned home very late that night and went to bed. She had been sleeping for barely three hours when the telephone jarred her awake. It was an aviation official from Mitchell Field, Long Island. The weather had finally cleared up, he said; would Elinor like to try for the endurance record at last?

She would. But she had to sneak out of the house before her parents noticed that she was awake, or else they would forbid her to go on such short notice. At twelve o'clock noon, on February 1, Elinor Smith took off in a Brunner-Winkle plane. She flew for several monotonous hours until suddenly she saw her father flying next to her in his own plane. He had come up with the message—written on the fuselage—that by eleven o'clock that night the moon would be out. This was a piece of luck, because every light on Smith's instrument panel had blinked out.

When Elinor Smith saw a searchlight coming from a radio station at Bellmore, Long Island, she flew toward it. Its beam illumined the clock on the plane's instrument panel: one o'clock in the morning. Elinor Smith had broken Trout's record by one hour and fifteen minutes. She landed happily and went home to catch up on her sleep.

But Bobby Trout was unwilling to let Elinor Smith keep the record. Twelve days later, Trout flew in a Golden Eagle for more than seventeen hours. The endurance records were coming fast. Louise Thaden, soon to become famous in airplane meets and races, stayed aloft for twenty-two hours, three minutes and twenty-eight seconds, flying in Oakland, California, in a Travelair. This was on March 18, 1929. Four new endurance records for women had been set since the New Year.

Elinor Smith was a stubborn pilot, however, and she wanted the endurance record back for herself. This time she had the aid of the Bellanca airplane manufacturers, who wanted her to break Charles Lindbergh's nonstop record of thirty-three and one-half hours. On April 24, Smith took off. The airplane was equipped with enough fuel to last for almost forty-eight hours, and it was filled with enough food and chewing gum to last Smith for the same amount of time. She also brought a book with her to make the time go faster—*Tom Sawyer Abroad*. Unfortunately, Smith was forced to return to earth after twenty-six hours, twenty-one minutes, and thirty-two seconds. This was through no fault of her own—a stabilizing cable in the plane had worked loose—but Smith was extremely disappointed. Though she had set a new women's record, she had not managed to beat Lindbergh.

So she returned to altitude flying for a while and broke her own altitude record when she flew 27,419 feet on March 10, 1930. Unfortunately, Smith was not allowed to keep this record for very long. Only a year later, Ruth Nichols took it away from her by flying over one thousand feet higher.

This was too much. Smith wanted the altitude record back for herself, and she decided to make another try for it. Her takeoff from Mitchell Field in March, 1931, was uneventful, but when the plane reached twenty-five thousand feet above New York City the engine suddenly died. Smith tried to stay calm. She kept her oxygen tube tightly between her teeth and looked for the source of the trouble. In her haste, she cut off her oxygen supply and blacked out completely. When she regained consciousness, the plane had dropped twenty thousand feet.

Smith realized that she would have to make a crash landing. She managed to grab off her goggles and to turn the plane over before it landed between two trees. The plane was ruined, but Smith crawled out of it completely unhurt and still determined to beat Ruth Nichols.

In two weeks she had recovered her courage enough to want to make another flight. Smith flew in a Bellanca Skyrocket monoplane and flew very well: the altimeter read thirty-two thousand five hundred feet when she returned to earth. Mr. Bellanca, the plane's designer, came to the field to offer his congratulations, newspapers rushed to prepare stories about the flight for their next issues, and Ruth Nichols telegraphed her congratulations. But all the excitement was for nothing. The altimeter, it was discovered, had been wrong. Elinor Smith's plane had actually gone less than twenty-five thousand feet high, and the world record still belonged to Ruth Nichols.

Ruth Nichols was born in 1901. Her father was a stockbroker and her mother a Wellesley alumna. The family was listed in the Social Register, a fact which—along with the fact that she was very photogenic—made Nichols seem even more romantic to the public. (She also wore a custom-made purple leather flying suit which must have helped attract public attention as well.) Nichols once said, "Records are made to be broken, and I only wish that more girls could get good ships and keep setting new marks all the time. It has long been my theory that if women could set up some records, in many cases duplicating the men's, the general public would have more confidence in aviation."

Nichols acted on this belief. She began flying early, while she was still a student at Wellesley. Her first flight was with the famous Eddie Stinson, Katherine's brother. As she recalled later, before that first flight she had been terrified even of elevators. After the flight—which Stinson livened up by looping the plane—she felt like a caterpillar who had suddenly become a butterfly. Two years before graduating from college, she made the first nonstop flight from New York City to Miami. In 1930 she made two transcontinental records, and one of them came very close to beating the men's record.

Oddly enough, Ruth Nichols did not yet have her basic pilot's license when she made those record flights. When she applied for her license, she must have been nervous, for her flight inspector remembered that three times she failed the basic test he had set her. "I'd look pretty stupid turning her down when she just set a speed record from L.A. to New York," he said in an interview. "She came into my office and shook her head and said, 'I can do better than that.' I sat down and gave her the rating."[1] On the other hand, Nichols passed perfectly the test for being licensed as a seaplane pilot, becoming the first woman in the United States licensed to pilot a seaplane.

In 1931 Ruth Nichols set the altitude record that was to cause Elinor Smith so much trouble. She decided to make this flight for a secret reason: she hoped to be the first woman ever to fly solo across the Atlantic. In 1930, Nichols had convinced the president of the Crosley Radio Corporation to lend her his plane so that she could set three women's world records. These, she felt, would help give her the stamina she needed to make a transoceanic flight. In Crosley's plane she set one transcontinental record. The flight for altitude was the second trial she set herself.

The first step in getting ready for the flight was making the plane as light as possible. To do this, Nichols took everything possible out of the plane: extra seats, seat padding, the pilot's backrest, and the door handles. She also hoped to be as light as possible herself when she made the flight, and accordingly she began dieting and exercising rigorously. The plane was equipped with new streamlined landing gear which required many test flights before she could feel comfortable with it.

Everything was finally ready on March 6. One newspaper headline that day read—perhaps with an unintentional morbidity—"Ruth Nichols Turns Angel Today." The day was clear but bitterly cold, and Nichols knew how much colder she would become at a high altitude. She wore four sweaters, long underwear, a flying suit made of reindeer leather, reindeer boots over two pairs of wool socks, reindeer mittens, a helmet snugly lined with fur, a scarf, and a parachute.

[1] Vecsey, George, and Dade, George C., *Getting off the Ground*, p. 267.

61

Even all those layers could not keep her warm enough. When the plane reached twenty thousand feet, the wing thermometer had gone as low as it could—forty-five degrees below zero. Nichols decided to turn on the oxygen she would need at such a height. (Oxygen use had become common in the mid-nineteen-twenties, after military testing; in planes of the era, oxygen was needed above a height of 28,000 feet.) As was customary in those days, she put the oxygen tube into her mouth. "As the plane climbed steadily higher," Nichols remembered in her autobiography, "I became more and more conscious of the intense cold—especially in my tongue. I was sucking oxygen directly from the steel tank in the wing, where it did not have even the slight benefit of the cabin heater. I tried moving my tongue; it seemed to be a solid chunk of ice. I removed the tube from my mouth for a moment, and this brought some relief, but I knew this was dangerous business at such a high altitude, for if I didn't maintain a steady flow of oxygen I might black out."[2]

Soon Nichols noticed that she was moving more and more slowly. Any motion at all made her feel faint and dizzy. She had been in the air for thirty-five minutes and had reached 28,743 feet, and she was jubilant. "I guess I was somewhat lightheaded from my erratic oxygen intake, my frozen tongue felt like a large ice cube, but what did I care? I was higher than any woman had ever flown before."[3]

The next year, Nichols set another altitude record, this one for diesel-powered airplanes, flying a Lockheed Vega at 19,928 feet. But people who set records can never relax. Maryse Bastié of France had begun to make a name for herself in aviation, and she was preparing to take Nichols's record away.

[2] Nichols, Ruth, *Wings for Life*, p. 124.
[3] Ibid., p. 125.

Like Elinor Smith, Maryse Bastié excelled at both endurance and altitude flying. (She was also skilled as a distance flyer.) Christened Marie-Louise Bombec, she had never planned to be a pilot as a child in Limoges; instead, she hoped to be an admiral. But the reality of her childhood was considerably more drab than her ambitions. Her father died when Maryse was ten, and her mother decided to put the little girl to work as a cobbler. When the first World War broke out, she worked as a seamstress, making overalls for doctors and nurses, and then decided to become a secretary so that she would have work when the war was over. In 1918 she married Louis Bastié, an aviator who was working as a shoe salesman. When he was appointed monitor at the Aviation School in Bordeaux, the couple moved there, and it was there that Bastié's life as a pilot began.

In 1925 Bastié received her flying certificate, and immediately she began wondering how to put it to use. As she said, "I couldn't just go on flying up and down the runway like the other pupils. I already felt that I simply had to attract public attention." Again like Elinor Smith, Bastié was sure she would get that attention if she flew under a bridge. She set out to do so a week after receiving her certificate. Bastié managed to keep clear of both the bridge and the water—not nearly as easy to do as it sounds—but when she emerged from under the bridge she realized that she was lost. To her embarrassment, she had to fly slowly above a tramcar in order to find her way back to the airfield.

A year later Bastié was forced to acknowledge what a serious career she had chosen for herself. Her husband was killed in a plane crash, and she did not fly herself for two years after his accident. She only decided to return to the air when a new school for women pilots opened up at Orly. Bastié joined the school as an instructor and shortly after that began to set endurance, altitude, and distance records.

63

Bastié managed to beat Elinor Smith's endurance record by less than half an hour when she talked the Caudron airplane manufacturers into lending her a plane with an extra fuel tank. In it she was able to stay aloft for twenty-six hours and forty-eight minutes. Lena Bernstein, also of France, beat this record at Le Bourget, France, on May 2, 1930, when she flew for thirty-five hours, forty-five minutes, and fifty-five seconds.

Bobby Trout had decided in the meantime to set a new kind of endurance record. She wanted to be the first civilian pilot to refuel a plane in midair. If the pilot did not have to land for refueling, endurance flights could become much longer. However, refueling in midair could not be done unless there were two pilots in the plane. Bobby Trout cannily asked Elinor Smith to fly with her. In December, 1929, the two women stayed together in the air for a remarkable forty-five hours and five minutes.

Back in France, Maryse Bastié was certain that she could beat Lena Bernstein's time without having to refuel. Once again she begged a Caudron for the flight. As officials warned her, the plane was virtually pure fuel tank. It weighed 880 pounds when empty; she was to fly it with 990 pounds of fuel.

The potential dangers such a flight posed did not bother Bastié at all. On the evening of September 2, 1930, she lifted into the air. For thirty hours she managed to fly fairly easily, although the smell of exhaust made her feel dizzy and sick. When she was into her second night in the plane, though, she began to become dangerously groggy. Weariness made her hallucinate: all around her Bastié saw planes swooping towards her "at extraordinary angles." When she realized that she was about to fall asleep despite all her concentration, she decided on a desperate remedy. She deliberately opened a bottle of cologne and splashed it into her eyes. "That wakes you up all right, I give you my word!"

Seeing that she was almost out of fuel came as a relief. Bastié was simply too tired to care about staying up longer. Now finally she could land. She had stayed up for thirty-seven hours and 55 minutes. Her eyes were watery and bloodshot and her hand was throbbing from holding the control stick, but Bastié had been in the air longer than any other solo pilot. Bastié now turned her efforts to altitude flying. When she broke Ruth Nichols's altitude in 1932, she flew 32,122.6 feet high. Two years later, on August 19, 1934, she broke her own record by only sixteen feet.

Another French aviator, also named Maryse, is, like Bastié, best remembered for her long-distance flights. Maryse Hilsz set several altitude records as well, however. One of these, made on August 19, 1932, was unofficial, which was a pity: Hilsz beat Ruth Nichols's official record by more than four thousand feet. Hilsz set another unofficial altitude record in 1935, flying up to 37,704 feet. The official record she established in 1936 was truly worth making official. Hilsz flew 46,947 feet high, smashing all men's and women's altitude records at the time. This record remained untouched for more than thirty years.

It seemed that altitude flying had become the special province of French aviators. Three altitude records were established for light planes in 1935. One of these was set by an Argentinian, Carolina Elena Lorenzini; the other two belonged to France. Maryse Bastié set one of the two. The third record for light planes was set by Madeleine Charnaux.

Unlike her countrywoman Marysé Bastié, Madeleine Charnaux came from a well-to-do family. Her brother, father, and grandfather were physicians, and her family paid for her flying lessons after Charnaux had abandoned sculpture for aviation. Charnaux created some commotion when she arrived for her lessons in a custom-made flying suit; usually only well-established pilots indulged in such frivolities.

But Charnaux was not just a rich girl playing at being a pilot. She had already seen how dangerous aviation could be. On her first day of lessons, she had made friends with another aviation student, Muguette Henry. Only two days later, Madeleine was asked to make a contribution towards a wreath for Mademoiselle Henry, who had died in a plane crash.

After setting her altitude record in 1935, Charnaux began to practice "blind" flying, in which the pilot relies completely on instruments for guiding the plane. Ill health and a back injury began to weaken her seriously, however, reaching a point where even landing a plane was agonizingly painful to her. Once the effort of standing at attention made her faint. Even so, she was able to set a few more records, but Charnaux knew she was dying. She asked permission to make one more flight over Paris during the war. It was the last time she flew.

The Russian aviator Poline Ossipenko set not one, but three women's altitude records on May 25, 1937. She flew in a seaplane—which could take off from and land on the water —to a height of 29,081 feet on her third attempt.

Back in the United States, endurance flying continued to be popular. Bobby Trout had decided to set another world's refueling record, and in January, 1931, she and another flyer, Edna May Cooper, flew together for 122 hours and twenty minutes. Astonishing as this figure was, it did not last long. Viola Gentry was now promoting a women's endurance contest, and she picked Louise Thaden and Frances Marsalis to try for a new refueling endurance record.

Texas-born Frances Marsalis had become an expert stunt pilot, but she had had no experience in endurance flying.

Her flight with Thaden certainly provided that experience. A male endurance pilot once remarked that tempers become uncontrollably short after thirty hours in the air. If that is true, Thaden and Marsalis must have had wonderful patience with each other. They began the flight in August, 1932, and did not return to earth until more than eight days had passed.

Evidently Frances Marsalis was a stoical woman, for in December, 1933, she made another refueling endurance flight. For this flight her companion was Helen Richey, who was then virtually unknown. It was a trying Christmas season for both women. The occupants of the refueling plane tried to make things merrier by dropping a turkey, a small imitation Christmas tree, and Christmas cards from fans to the two women. But Marsalis also had to celebrate her birthday aloft, and by then both she and Richey were too tired to want any celebration. (Marsalis dropped a forlorn note down to announce, "I'm growing a year older every day I'm up here.") Richey and Marsalis finally returned to earth after nine days, twenty-one hours, and forty-two minutes in the holiday skies. It took them a long time to become rested enough to appreciate their great accomplishment.

Grace Huntington set two national altitude records in a light plane, one in 1939, and one in 1940. When World War II broke out and United States civilians were no longer permitted to fly, she was the only woman in the country to hold any light plane records. Huntington hoped that the two records would be proof of her ability to test planes during the war, but few airplane manufacturers took her seriously. Instead one company hired her brother, whom Huntington had had to coax into flying at all and who had had much less practice than she.

Grace Huntington was not discouraged. She set her hopes on learning more about altitude flying, saying, "I just scraped the bottom of the top. There's a lot to learn about the upper air. I hope I play a part in getting it. I want a little recognition—not for myself, but for all the women who fly—which will result in jobs which we know we can fill."[4] The recognition Huntington and other record-setters brought was one of the most crucial contributions to women's aviation before World War II.

[4] Planck, Charles E., *Women with Wings*, p. 106.

Over Mountains, Deserts, and Oceans

A LL THE MEN learning to fly at the Caudron Factory in France agreed that Adrienne Bolland was one of the most infuriating students there. The year was 1919, the place Issy-les-Moulineaux, and the plane Adrienne was learning how to fly was a Caudron G-3.

The G-3 was a clumsy old plane, difficult to fly and fond of spraying oil over the pilot. Moreover, Bolland first went up in it without any flying experience at all; she had signed up to be trained for her pilot's certificate and had paid her two thousand francs to Caudron without ever having been up in a plane.

Bolland had a tough time learning how to fly at Issy-les-Moulineaux. Her instructors disliked having to teach women. When she told one airline manufacturer that a plane of his was unreliable, he told her not to use it any more. He was willing to believe her, he said, but he thought it would cause less trouble to have a man killed in the plane than a woman. Nor were her fellow students receptive to the idea of studying with a woman. They criticized Bolland constantly; in return, she kicked and slapped them.

But she was nonetheless learning how to fly. Monsieur Caudron promised her an airplane to use for her own if she would loop the loop for displays. Once she had the plane, Bolland asked Caudron for permission to take it on a long-distance flight. He thought she was joking, so he jokingly suggested that she fly over the Andes, a feat no one had done successfully before. Bolland agreed immediately and began to prepare for the flight before Caudron had the chance to take back his offer.

On April 1, 1921, Adrienne Bolland set out from her hotel in Mendoza, Argentina. "I kept up my high spirits, and was the life and soul of the party," she recalled, "though the others gazed at me with a kind of pity. It was as though they were attending a funeral and the corpse, by way of a joke, was taking its time and holding everything up, playing all sorts of tricks in the most frightful bad taste. I could see from the looks of them that they were all depressed and already mourning for me."[1]

She herself felt much more confident. When the plane was in the air, though, it took Bolland a long time to climb high enough to clear even the lowest mountain in the range. Finally she decided to stop trying to climb and instead to fly the plane alongside one of the mountains. Once she had started flying parallel with the mountain, the wind began to lift the plane very gradually. "Suddenly I saw an opening between the hilltops, with a patch of blue about one thou-

[1] Lauwick, Hervé, *Heroines of the Sky*, p. 40.

70

sand feet down and sloping ground behind it. Beyond, in the distance, straight ahead, lay the plain of Chile. I was through!"[2]

Bolland's success created enormous excitement in the aviation world, especially in France. For one thing, the flight was extremely useful a few years later when the French Government approached the governments of Brazil, Argentina, and Chile in order to create airlines between Europe and South America. More important, the flight inspired several of the greatest women aviators to make some of the most famous flights in aviation history.

The idea of a long-distance flight seems especially challenging. It demands tremendous physical stamina on the part of the pilot, who may be forced to land far from any human help and who must be completely self-reliant. It also demands an extremely serviceable plane.

Distance flights only became possible toward the end of the 1920s, a decade that saw many improvements in airplane design. Planes were now made of metal; metal propellers, enclosed cabins, retractible landing gear (which could be drawn up into the plane when in flight), and air-cooled engines (which were lighter than the old water-cooled ones) were beginning to be standard equipment. These and other improvements made longer, more strenuous flights possible.

One of the first women to make an exceptional distance flight did so in 1927. Mrs. Keith Miller of England and her co-pilot, Captain W. N. Lancaster, set out from London on October 15. Their plan was to fly to Port Darwin, Australia, a distance of twelve thousand five hundred miles. Miller's and Lancaster's plane broke down several times on the way. Their first crash was in an Asian desert, but the plane required only simple repairs which the pilots could make themselves. The next crisis came in the skies above

[2] Ibid., p. 40.

Burma, when Lancaster spotted a poisonous snake under one of the plane's seats. He tried to kill it, but the snake escaped into the cabin where Mrs. Miller was at the controls. She had to fend it off with a long pole and try to fly at the same time. Finally she succeeded in killing the snake and they continued on.

But the two pilots were not yet free of misfortune. On January 10, their plane crashed in Indonesia, injuring both of them and breaking Mrs. Miller's nose. After this crash they were forced to wait while the plane was repaired in Singapore. Despite these hardships, Mrs. Miller and Captain Lancaster did arrive in Port Darwin on March 19, 1928. No woman had ever flown as far before.

The "Lady Marys" of England both made flights from London to Cape Town, South Africa, in 1928. Lady Mary Heath flew eight thousand miles from London to Cape Town, beginning on February 12, and arriving on May 17. Mary Heath was a picturesque traveler. She steadfastly refused to wear a sensible aviation suit when she flew, feeling that unfeminine clothes cheapened women's flying. Instead she preferred a dress and high heels. She also insisted that her plane be painted an intense turquoise so that it would match one of her favorite rings.

There was nothing picturesque about her flight. She flew completely alone, except when crossing the Sudan; in Cairo, British officials had insisted that she take a passenger for that most dangerous section of the flight. They could not protect Lady Heath from the African heat, however, and she was hit by sunstroke over Rhodesia. She almost fainted and was forced to make a crash landing. When she had recovered, Lady Heath gamely continued on to Cape Town—the first woman to have flown there successfully from London. She also managed to set a speed record from Cape Town to Cairo on the return flight to England.

The African flight firmly established her credentials, and

on July 28, 1928, Lady Heath was made a Royal Dutch Airlines copilot on an Amsterdam-to-London flight. This honor made her the first woman to act as a pilot on a passenger plane.

Lady Mary Bailey had been active in the meantime, flying a DeHaviland Moth round-trip from London to Cape Town. She left on March 9, and arrived almost two months later on April 30. Lady Bailey returned to London by way of the Belgian Congo, a distance of ten thousand miles. Much of her flying was done at night so as to escape the African heat. Lady Heath had been the first woman to fly the whole length of Africa; Lady Bailey became the first woman to do so completely alone.

A third British peer to make an important distance flight was the Honorable Mrs. Victor Bruce. Mary Bruce had already broken several speed records on land and sea; she was fond of car racing and also of speedboating and had set a new record by making a round trip across the English Channel in an outboard boat. (She had also been the first British woman—or girl, rather, since she was fifteen at the time—arrested for speeding after going sixty-seven miles per hour on her brother's motorcycle.) Her love of adventure was well known, but some people thought that mere adventurousness did not justify Mrs. Bruce's plan in 1930: she intended to fly solo to Japan in a newly purchased Bluebird, though she had never flown in her life.

After buying her plane on a sudden whim, Mrs. Bruce called the British Air Ministry in June and learned that she would have to set out before October in order to escape the monsoons in India. So she had three months in which to learn how to fly. She bought maps (which the salesman made her pay for in advance when she told him her plans), mapped out her route, and had an extra fuel tank set up in the spare seat of the Bluebird. When all these preparations were done, there were only eight weeks left in which to

73

learn to fly. By the end of the first week Mrs. Bruce had soloed. At the same time she took a crash course in navigation, which included learning how to use a compass.

In the early hours of September 25, Mrs. Bruce took off from the Heston Aerodrome. Leaving the aerodrome was a first in itself, for she had never flown further than three miles from it before.

After four uneventful days of flying from sunup to sundown, Mrs. Bruce had arrived at the Persian Gulf. Here an oil pipe in the plane burst when she was crossing the Bay of Bandar Abbas, and she was forced to make a crash landing in the desert. Natives of the region took her message to Jask, Iran, and protected her until help arrived.

After the engine had been repaired, Mrs. Bruce set off again. Soon she was above the immense, monotonous jungles of Southeast Asia. But her delay at Jask meant that she was caught in the Chinese monsoon. While waiting in Siam for a clear day, she caught malaria. In all, it took Mrs. Bruce until the end of November to reach Japan. After this voyage she was part of the first air-to-air refueling flights in England and then did a stint with a flying circus. And after that she took first place in the Open Jumping class at the Royal Horse Show.

It must have seemed to aviators of the 1930s that Maryse Bastié could do anything. In 1931 she flew a small plane from Le Bourget, Paris to Urina, Russia and set a new official distance record of one thousand eight hundred fifty miles: it was the longest straight-run distance flown by a woman and the longest straight run ever flown by either sex in such a light airplane. Bastié said later that she had looked so bedraggled on alighting from the plane that when she asked where she could send a telegram she was immediately taken to the hospital. She added, "It's true, of course, that my Russian may have been less than perfect!"

Bastié's record was stolen three months later by the indefatigable Ruth Nichols, who flew nonstop 1,977 miles from Oakland, California, to Louisville, Kentucky, on October 26. This record was taken in turn by Amelia Earhart's transcontinental flight on August 25, 1932. Like Nichols, Earhart flew in a Lockheed Vega, and the distance she covered was 2,478 miles.

The next year, Maryse Hilsz flew solo twenty-five thousand miles from France to Indochina and then to Japan. The flight took six days and twenty-three hours. When Hilsz repeated the journey in January, 1934, she pared the time it took down to five days and ten hours. Hilsz's long-distance flights were considered a double victory for France. Not only did they bring world attention to French aviation, but they also improved France's relations with Southeastern Asia, with Hilsz serving as a sort of informal ambassador between the two.

For some reason, few long-distance flights involved South America. Laura Ingalls broke with tradition in 1934 when she toured South America by plane. Ingalls was already celebrated as a stunt pilot—she had looped the loop 980 times in a row, at a dollar a loop—and as a transcontinental pilot; she had been the first woman to fly nonstop from New York to California. Unfortunately her nonstop flight had taken longer than Amelia Earhart's previous east-west transcontinental flight, and Earhart had made stops on the way. Ingall's seventeen-thousand-mile South American circuit was unquestionably first-class, however, as well as being the first such flight made by a woman. Ingalls received the 1934 Harman Trophy in recognition of the flight.

The same year saw Anne Morrow Lindbergh presented by the National Geographic Society with the Hubbard gold medal. Lindbergh was a quiet, somewhat bookish woman, and her accomplishments as an aviator were often overshadowed by those of her husband Charles, but she was an extremely good pilot and navigator herself. After all, Charles

Lindbergh had taught her to fly. Anne Lindbergh made several long flights with her husband, serving both as navigator and as copilot. They made a survey of Central and South America for Pan-American Airways and then returned unofficially to see whether they could spot the ruins of any ancient South American civilizations from the air. The Lindberghs did in fact find one ruin, the remote Temple of the Warriors, almost invisible in the Guatemalan rainforest. In 1933 they embarked on a twenty-nine-thousand-mile aerial survey flight, and it was for this flight that Anne Lindbergh received the Hubbard medal.

Any long-distance flight is a challenging one, but ocean crossings in airplanes hold a special place in aviation history. If a plane should happen to break down above land, there is at least a chance that it may be brought safely to earth. But if it begins to falter above the ocean, there is of course far less chance of its coming down safely. Even someone riding in a passenger plane knows how different it feels to fly above water compared to flying above solid ground. For this reason, the stories of airplane voyages across the Atlantic—and the pilots who fly those airplanes—have always fascinated the public.

Many, many women aviators attempted to fly the Atlantic before one finally succeeded. In 1927, when her plane came down into the water 350 miles from the Azores, Ruth Elder had the dubious distinction of being the first woman ever to be rescued on a transatlantic flight. The German aviator Thea Rasche was in the process of preparing to fly across the Atlantic in 1928 when the airplane company supporting her withdrew its backing. In 1931, Ruth Nichols took off from Newfoundland fully intending to reach Europe; she had picked Newfoundland as a take-off point because it was closer to Europe than was any spot in the United States. That fact did not help her much, however, as she crashed into a hill immediately after leaving the airfield.

As everyone knows, it was the great aviator Amelia Earhart who became the first woman to fly the Atlantic successfully, both as a passenger and as a solo pilot. It is still Earhart whom most people think of when they think of women aviators, and with good reason. If any aviator deserves to be remembered, it is she.

It was to be expected that Amelia Earhart enjoyed a remarkable childhood—remarkable in the diversity of things she enjoyed, and remarkable for the amount of freedom she and her sister were allowed by their understanding parents. "Like all horrid children, I loved school," Earhart said, and she spent many hours reading as a child. Perhaps books increased her love of adventures, but certainly much of Amelia's childhood was also spent actively pursuing excitement. She was born in 1898, in a era when exercise was considered unnecessary for little girls, and she had to teach herself basketball, bicycling, tennis, and riding. She tobogganed off roofs and under horses; she built traps to catch the neighbors' chickens; she wore the first gym suit her home town of Atchison, Kansas had ever seen. The town had probably seen few girls who asked for footballs for Christmas, either. (Amelia asked for the footballs, while her sister Muriel asked for—and got—a popgun.)

After graduating from high school, Earhart decided to become a nurse's aid in Toronto. World War I had made her feel that further school was frivolous when she could be so helpful in veterans' hospitals. Her war work was what introduced her to aviation: she spent as much of her spare time as possible watching officers learn to fly. But nursing also introduced her to the world of medicine, and after the Armistice she went to Columbia University intending to become a doctor. Earhart soon moved out to California to be with her family. There, her interest in medicine did not last: "Somehow, I did not get into the swing of the western universities before aviation caught me."

Earhart's first airplane ride was with Frank Hawkes, a famous speed pilot. The minute she was up in the air, she realized she would have to learn flying. That evening she told her family she wanted to fly, "knowing full well I'd die if I didn't." Her parents gave her their permission, but no money. Earhart had to earn the money for her lessons herself.

Her first teacher was a woman—Neta Snook. "Snooky" was a generous teacher, generous enough to turn Earhart over to another teacher when she felt Earhart could learn more that way. It was under his tutelage that Earhart first soloed in 1920. She had expected to be nervous during the solo flight; she found, though, that she enjoyed all of it. She flew up five thousand feet—unusually high for a first solo—swooped around for a little while, and came down. Her landing was rough, but she had not been afraid to land as many novices are. Inexperienced as she was, it was clear from the solo that Earhart knew how to handle herself in a plane. Her mother was so impressed that Earhart had come this far that she helped her daughter buy a plane.

The practice she gained from owning her own airplane helped Earhart to set her first record for altitude, flying up to fourteen thousand feet. But except for this record, her flying career for the next few years was relatively quiet, for Earhart still needed to support herself and could not afford to fly much. Not until 1928 did the opportunity she needed come along.

It came by telephone one afternoon in April. A British woman named Amy Phipps Guest had bought an airplane from Admiral Richard Byrd. She had named it the *Friendship* and planned to fly the Atlantic in it from west to east. Her family had dissuaded her from making the flight, but she wanted an American woman to be the first to take her place. If Earhart would promise not to ask for pay for the flight and not to hold anyone responsible in the case of an accident, she was welcome to substitute for Mrs. Guest.

Of course Earhart was willing to go, and she hoped to be able to do some piloting on the flight herself. But she could not let anyone know she was going to make the trip. To do so would have brought too much publicity to the venture. The pilots sent out word that the plane was being prepared for a trip to the South Pole. This, they hoped, would explain to any curious onlookers why so much commotion seemed to be going on at the Boston Airport. Still, Earhart could not go near the *Friendship* herself. In fact, she had only one glimpse of the plane before the day it departed from Newfoundland. Not even her family knew what she was planning.

On June 17, 1928, the *Friendship* left Newfoundland. Aboard were pilot Wilmer Stultz, navigator Lou Gordon, and Amelia Earhart. The plane had been painted bright gold so that it would show up well if it had to land in the water. The three travellers had packed scrambled egg sandwiches, coffee, oranges, malted milk tablets, chocolate, water, and pemmican. However, they had not been able to plan the weather as well as they had planned everything else. Most of the trip was foggy.

Because of the weather, Earhart never had the chance to do any piloting. Instead she kept a diary of the flight and observed Bill Stultz closely, hoping to pick up tips that might be helpful later. She had plenty to watch: the plane's radio had worked only for the first part of the flight, so Stultz could not receive messages from ships to let him know where he was. All he had to use was the *Friendship*'s instrument panel.

After a day and a night, the travelers saw land. At first they thought it was what Earhart called "shadow geography" —clouds. But the land was real, and it was the coast of

Wales. It had taken twenty hours and forty minutes to cross the Atlantic, and Amelia Earhart had become the world's new heroine.

Her only regrets about the flight were that she had not been able to do any flying herself, and that Stultz and Gordon were generally ignored in the rush of publicity surrounding the trip. She felt that they, not she, were the ones who deserved the credit, having done the actual flying and navigating themselves.

In 1932 Earhart got the chance to earn the credit she had been given in advance. Again she kept her plans secret, again she kept away from the airport while her plane, a red Lockheed Vega, was being prepared. It was much harder to be secretive this time. Earhart was now a famous pilot, and she was going to become the first woman to fly the Atlantic solo.

Earhart left Newfoundland at 7:12 in the evening on May 20. The first few hours of the trip went fine, and after that almost everything possible went wrong. First the plane's altimeter broke, making it impossible for Earhart to tell how far above the ocean she was. Next she flew directly into a storm and narrowly missed being struck by lightning.

She decided she might be able to escape the storm by flying above it, "so I climbed for half an hour when suddenly I realized I was picking up ice." The rain from the storm had iced over the plane's wings as Earhart had flown higher. She would have to fly down again, back into the storm. The plane rattled and shook as it was buffeted by the winds, and this continued all night. Earhart was lightheaded from exhaustion and hunger: all she consumed on the flight was a can of tomato juice. In addition, she had noticed flames gradually eating away at a broken weld in the engine manifold. "The metal was very heavy," she wrote, "and I hoped it would last until I reached land. I was indeed sorry that I had looked at the break at all because the flames appeared so much worse at night than they did in the day-time."[3] To top everything off, the gas gauge in the plane's reserve gas tank was leaking, so there was no accurate way to tell how much fuel was left.

Finally, after fifteen hours and eighteen minutes, she reached Ireland. After frightening "all the cattle in the county," she landed in a Londonderry meadow.

"There ended the flight and my happy adventure." But it was by no means the end of her shining career.

THE BRITISH TOO had their "Queen of the Air," Amy Johnson. Amy Johnson was an excellent aviator and enjoyed flying, but her flying career always had a hint of sadness attached to it. Even her quiet dedication seemed tinged with sadness: "I shall fly till I die, and I hope to die flying."

It was rumored by the British press that Amy Johnson began flying in 1928 to give herself something to do after an unhappy love affair. Whether or not that was true, it was certainly true that Johnson gave herself wholeheartedly to learning to fly, paying for her lessons by doing secretarial work. She was an excellent student; her only problem was rough landings.

By 1930, Amy Johnson had her pilot's license and fifty hours of flying time. Those two things seemed like scant security, considering what she planned to do: she wanted to fly solo from England to Australia. A third measure of security came when Johnson learned how to fix her airplane, for she knew how difficult it might be to find help if she were forced to land in a remote spot during the voyage. In fact, Johnson took and passed the tests for a ground engineer's license, becoming the first woman ever to do so.

[3] Earhart, Amelia, *The Fun of It*, p. 216.

She also took along a letter assuring any potential bandits that the British government would pay her ransom if she were kidnapped.

The flight took nineteen days. Johnson flew twelve hours a day and then spent four or five more hours preparing her plane for the following day. In Asia, she had to combat the seasonal monsoons, and in Burma, on the playing-field of a boys' school, she made such a poor landing that her plane rolled into a ditch. (Johnson patched the plane's wings with some of the boys' shirts.) When finally she reached Darwin, Australia, Johnson again damaged the plane as she landed, but this time she did not mind: she had succeeded.

While her plane was being repaired, Amy rode with Jim Mollison, a commercial pilot, to Sydney, and they married soon after meeting. Mollison set a record himself a year later, flying from Lympne, England, to Cape Town, South Africa, in four days, seventeen hours, and fifty minutes. A few months after that, his bride broke his record—under exactly the same conditions—by more than ten hours.

Mollison seemed determined to outfly his wife. In August, 1932, he soloed the North Atlantic, and in 1933, he soloed the South Atlantic, becoming the first aviator to make an east-to-west transatlantic flight across that part of the ocean. At this point, the Mollisons decided to start flying together so that they would no longer have to compete against each other.

The two pilots felt another transatlantic flight, from Wales to New York City, would be an appropriate way to begin their joint career. They set out on July 22, 1933, taking turns at the controls. Storms and fog came up when they were midway across the ocean, and they flew into the wind for much of the voyage. This meant that the plane needed a huge amount of fuel, and once they reached the United States, the Mollisons, desperately tired and dangerously low on fuel, had no choice but to land before they reached New York.

At Bridgeport, Connecticut, they found a suitable airfield and prepared to land. James Mollison was so tired that he could not approach the field correctly; the plane came down in a swamp. Mollison was knocked unconscious, but Amy Johnson remained conscious, and she knew she had to free her husband from the overturned plane before it caught on fire. She managed to drag him away to safety although her left arm had been badly hurt. The trip had ended unhappily. Still, Johnson had become the first woman to fly the Atlantic from east to west.

Soon after this flight, the Mollisons separated. Jim Mollison began to work as a consultant for another woman aviator, Beryl Markham, who became the first woman to make a solo east-west transatlantic flight. Markham's flight was all the more remarkable because it was made "blind"; she relied almost totally on instruments to guide the plane.

Amy Johnson quietly returned to solo flying. She managed to set several long-distance speed records before her death in the second World War.

Another woman who was often compared to Amelia Earhart was New Zealander Jean Batten. Batten was not only compared with Earhart; in 1936 the two women shared the award of the Harmon Trophy for Women, the highest award of the International League of Aviators.

Someday, someone should determine the percentage of women who have given up careers in the arts to pursue careers in aviation. Jean Batten was one more of these, selling her piano to pay for her plane lessons. At the age of nineteen, with her mother, Batten left New Zealand for London, where she began learning to fly in earnest. By the time she was twenty-three, she had a commercial pilot's certificate and she had learned a great deal about airplane mechanics. But simply being talented was not enough to keep her in the air. Batten knew that she could not afford to continue flying unless she made herself famous.

Being persuasive was another of Jean Batten's talents. She persuaded Lord Wakefield, an oil millionaire, to back her flight if she would use only Wakefield fuel in the plane. With his support, she took off on May 6, 1934, to fly from England to Australia. The trip took her only fourteen days and twenty-two hours; she beat Amy Johnson's record by four days.

The trip to Australia gave Batten's career the boost it had needed. She earned enough money by lecturing to pay back all her debts, and she found that she had also earned a reliable reputation: she was promised a government grant for whatever her next trip might be.

And Batten knew she would need that grant, for it was her secret plan to fly alone for one thousand eight hundred sixty miles, starting off from Lympne and flying to Senegal; from there she would fly across the South Atlantic to Natal, Brazil. Only James Mollison had flown this route before. Batten hoped to be the first woman to do so. She hoped as well to beat Mollison's record—and to do it without a radio in her plane.

Batten knew that she could beat Mollison only if she did not rest at all when she stopped. She took off at six o'clock in the morning on November 11, 1935, and flew to North Africa. There she spent one night before leaving Dakar. It was hard for her to nerve herself to leave that day. The Atlantic was stormy, and it was raining so heavily that she would be able to use only her compass for direction. Batten circled indecisively for some time, hoping that the rain would stop pounding the plane. Finally she took a deep breath and set off.

The rain never stopped for eight hours. Winds shook the plane, and a bolt of lightning broke the indicator that measured engine conditions. Batten had been tired even before setting out from Dakar. Now she was totally exhausted. Fatigue made her emotions erratic, and she burst into tears as she flew. But at the end of thirteen hours, she was rewarded by the sight of land. Her navigation had been perfect, despite the storms. Batten had flown from England to Brazil in only sixty-one hours, beating James Mollison's time by a day.

All her weariness vanished. She was so excited, she said, that she "burst out singing and thumping the wall of the cockpit as if it were the flank of a gallant steed."[4]

Meanwhile, Maryse Bastié of France had decided to solo the South Atlantic—or, as she put it, "that Atlantic of theirs"—herself. Having pumped Jean Batten for details, Bastié was confident that she could accomplish the flight, but few others were. Lefèvre, one of the first men ever to cross the Atlantic, remarked coolly that since Bastié was neither very rich nor very pretty, she would certainly never be able to cross it herself. Nevertheless, on December 30, 1936, Bastié took off from Dakar in a Caudron Simoun. She landed in Natal, Brazil, twelve hours and five minutes later.

"I kept banging away on the box and shouting, 'Here we are, then, here we are! It's Natal, it's Natal.' "[5]

While she was in South America, Maryse Bastié flew solo across the Andes. It was her tribute to Adrienne Bolland, who had risked that first dangerous flight fifteen years before.

One aviation writer remarked that all books on women aviators seem to take on the quality of a cemetary, packed with crosses, and it is unavoidably true that many of the most brilliant aviation careers ended in death. In 1937 came what is probably the world's most famous flight, and it was

[4] Jablonski, Edward, *Ladybirds: Women in Aviation*, p. 117.
[5] Lauwick, Hervé, *Heroines of the Sky*, p. 119.

never completed. Once more the pilot was Amelia Earhart, and her fate was never discovered.

Earhart's career was very carefully managed by her husband, George Putnam. He was a skilled businessman and a clever promoter, and there is no doubt that he helped advance her career. But it was said as well that he pushed her too hard into making flights she was not capable of making. Perhaps Earhart was not ready to fly around the world in 1937 when she tried to do so.

Still, Amelia Earhart herself was completely aware of the dangers ahead of her, and she faced them gladly. What she wanted to do was to fly around the world at the equator—the longest way around. If she succeeded, she would become the first woman ever to fly around the world and the first aviator, either male or female, to fly around the equator. She planned the flight to be her last before retirement, announcing, "I feel that I've got one more big flight in me."

Earhart would have preferred to fly alone, but she had to take along a navigator because the flight was so dangerous. The man she chose was Fred Noonan, who had been married for only a month before the flight. They took off from Oakland on May 21, planning to fly east to avoid flying into the wind for too much of the flight. Flying east meant that the most dangerous part of the voyage would come at the end.

Everything went smoothly until the plane reached Africa, where Earhart ignored Noonan's advice and flew in the wrong direction for many wasted miles. After righting her direction, she flew without trouble across Asia. Twice Earhart and Noonan were forced to land after reaching Indonesia: the plane's navigation instruments were faulty. After repairs, the two continued on to New Guinea. There they discovered that the plane's radio was also beginning to fail, but they were too impatient to wait for it to be repaired.

The most difficult part of the trip lay ahead, and Earhart wanted to get it out of the way.

She was heading from New Guinea to Howland Island, which lay 2,556 miles from New Guinea—a distance over the ocean that no pilot had ever tried to fly. In addition to its dangerous distance from New Guinea, Howland Island was also dangerously small, only one-and-a-half miles long and one-half mile wide and less than twenty feet above sea level. Moreover, its runway was so new that no one had ever landed a plane there before. In fact, the entire flight would have been completely impossible without the Coast Guard cutter "Itasca" to stand by the island and transmit help by radio. But the Itasca could send signals for only the last 500 miles of the trip; the plane's radio only had a 500-mile range. For one thousand five hundred miles Earhart and Noonan would be cut off from all communication.

The two took off at ten o'clock in the morning. They had enough fuel to last for twenty-four hours, though they hoped the trip would last for only twenty.

Not until 2:45 in the morning did the Itasca hear Amelia Earhart's voice. They could understand nothing except the words "cloudy and overcast." At 6:15 in the morning Earhart's voice finally became understandable. She reported, "We are about a hundred miles out. Please take a bearing on us and report in half an hour. I will transmit into the microphone." She whistled to give the Itasca a signal by which to check her location, but the static was too strong for the radioman to be able to do this.

Except for one brief message, the Itasca did not hear Earhart's voice again for an hour and a half. By seven in the morning, the ship's crew had gathered to watch the two pilots arrive, but they never came. At 7:42 Earhart's voice sounded clearly over the radio: "We must be right on top of you, but we can't see you. Our gas is running low. Have been unable to reach you by radio. We are flying at an altitude of a thousand feet. Please take a bearing." Again this could not be done; it is very difficult to locate an airplane using only the pilot's voice, and static made it impossible.

At 8:45, Earhart reported, "We are running north and south. We have only an hour's fuel left and we cannot see land." Desperately the ship's officers called her; frantically the men on deck searched the skies. They knew that the plane's fuel would be exhausted by ten o'clock. At noon they gave up.

Amelia Earhart and Fred Noonan were never found. The United States Navy organized the largest air and sea search in its history, combing an area of 265,000 square miles for sixteen days without discovering a trace of the plane or of its two occupants. Various explanations for their disappearance have been offered over the years; none have ever been proved. The world took what consolation it could from Earhart's own words, spoken just before her last flight: "I want to do it because I want to do it. Women must try to do things as men have tried. When they fail, their failure must be but a challenge to others."

The Record-Setters —
Speed

O N July 11, 1949, a small French plane carrying four people crashed into the Seine. One of its passengers was a woman named Jacqueline Auriol, the daughter-in-law of France's president and a skilled pilot in her own right. Of the four people in the plane, she was by far the most seriously hurt in the accident.

Auriol's injuries were terrible. As she lay in the river waiting to be rescued, she realized that she had lost all her teeth; once in the hospital she was told that, among other things, she had three skull fractures, that both jawbones and several ribs were broken, and, worst of all, that her face had come detached from her skull. Her first request, made on

the way to the hospital, was for a plastic surgeon. The next
thing she asked—and she asked it repeatedly—was whether
she would be able to fly again soon.

It was only when she was allowed to fly again that
Jacqueline Auriol really began to recover. From then on
her improvement was rapid. Less than two years after the
accident, flying in one of the first jet planes, Auriol set a
new world speed record. Over the next twelve years she
would manage to set four more.

The women who broke aviation speed records needed
what Auriol described as a special inner force. In her autobiography, she explained, "The prospect of breaking a
record has always plunged me into a state of excitement
comparable to nothing else. Every time I think of it, I feel
my strength multiplied a hundred times, like Popeye when
he's eating spinach . . . I really become another person."[1]
Record-holders, she said, are all linked in one way: they
are unable to resist challenge, challenge that comes both
from within themselves and from the outside world.

From the earliest days of aviation, the women who broke
speed records have had that quality in common. Jacqueline
Auriol was one of the greatest women speed flyers in the
history of flight, but she was by no means the first.

Like many other women aviators, Hélène Boucher never
grew up planning to fly. She chose flight as a career only
after realizing that she could not make herself become
interested in the various careers expected of her. Boucher
was born in Paris in 1908, the same year that the first woman
in history ever rode in an airplane. The daughter of an
architect, Boucher was raised conventionally; no one looking at her would have guessed her to be any different from
other well-bred children. But when she turned sixteen,
Boucher began to change. Suddenly she lost interest in

---

[1] Auriol, Jacqueline, *Vivre et Voler*, p. 207 (author's translation).

school and in the everyday life for which she was being prepared. Almost everyone feels this way at some point, but few people do much about it; Boucher decided to act.

She left school and left home, moving to England for several months. On her return to Paris, still unsure of what she wanted to do, Boucher tried several things: art, music, and finally—somewhat anticlimactically—dressmaking. Soon she had advanced as far as to become the manager of a clothing shop. It was during this unchallenging stage that Boucher met a reserve pilot who first introduced her to flight. On the fourth of July, 1930, Boucher flew as a passenger at Orly for the first time. The short flight changed her life. The excitement and speed of flying were things that stirred her whole imagination. (So too was the risk; a young pilot she knew had been killed in a plane crash.) Here was a career that could not possibly be dull. From then on, Boucher thought only of learning to fly—and of flying her own plane. She spent all her spare time at Orly, watching and learning as much as she could, and soon she caught the attention of the man who first helped her achieve her goal.

This was Henri Farbes, whose own goal was to open a flying school. Would Boucher like to be his first pupil? he asked. She certainly would, and she began studying with him in March, 1931. Later Farbes recalled her as one of the best students he had ever had. She pushed herself constantly, she asked questions all the time, and she was never satisfied with herself although she had become expert almost instantly. Within three months she had received her tourist's license and about one month after that, her public transportation license.

Boucher was beginning to attract attention at this point, but her first real fame came, oddly enough, after her first real failure. In 1933 she planned a long-distance flight from Paris to Saigon, and she left Paris on February 13. Just before she reached Bagdad, though, a crack appeared in the plane's crank-case, and Boucher and the plane were forced to land, almost hitting some donkeys in the process. She asked Air-Orient to help her; officials sent one hundred reluctant natives of the region, who loaded the plane onto a truck and drove it to Bagdad, getting lost twice on the way. When they finally did reach Bagdad, the damage to the plane was too serious for Boucher to be able to continue. She flew back to Paris after six weeks, humiliated by her lack of success, and found on her return that she had become famous. At the age of twenty-five, Hélène Boucher was a French idol.

She had by this time begun to be noticed by an engineer who had been following her progress for over a year and had decided that Boucher was unquestionably better than almost all existing male test pilots. His firm—Caudron Renault—was looking for a test pilot and decided to sign her on. It was a daring decision on their part and one that paid off. On July 8, 1934, Boucher took command of a Caudron Rafale in the twelve-hour competition flight at Angers. She made good time, 235 kilometers per hour, and placed second out of ten competitors.

Boucher did not rest there. Following her success at Angers, she decided to try secretly to raise the women's speed record in her Rafale. On August 8, she succeeded, flying an average of 409 kilometers per hour over a one-hundred-kilometers course. This beat the record for all categories, but still Boucher was not content: she wanted the record for pure speed. Only two days later she reached this new goal, flying 428 kilometers per hour. By now she was completely consumed by the need to fly even faster. On the next day, August 11, she made yet another try and flew 445 kilometers per hour, an amazing improvement over the previous day's time.

It had been an incredible August. Boucher's flight on

August 11 set a new world record in every category. She was not only the fastest woman aviator in the world; she was the fastest aviator known. The victory belonged not only to her but to France—and the French, who have always loved skilled pilots, responded ecstatically.

They did not have much longer to appreciate her. Boucher had always been superstitious about the number thirty; she swore that it was unlucky for her, and she was certain that she would never reach the age of thirty herself. Whether or not her superstition was well-founded, the fact that she did not get much practice in flying for several months after August made Boucher's next flight—on November 30—much less skillful than her record flights in August had been. As she was beginning her approach to the landing field at Guyancourt, Boucher miscalculated and missed the field. She came in to make another try, but again she judged the distance wrong, and this time she brushed some shrubbery bordering the field. When she tried to lift the Rafale out of the bushes, it flipped over and crashed to the ground. Boucher was killed instantly.

Hélène Boucher's accomplishments have never been forgotten. Her unswerving allegiance to flight earned her the cross of the Legion of Honor—at that time an unheard-of award for a woman—and a later award was established in her honor. Boucher had always been quiet about her achievements, and she would probably not have made much fuss about these awards, but they were fitting tributes to her. Her work had completely changed the role of women in aviation.

An American who was to have a similarly important effect on women's aviation also began to make herself known in the 1930s. She was Jacqueline Cochran, and she embodied the American success story. If it is true that women aviators are drawn by challenges, then Cochran was born into the ideal set of conditions to make her a superb pilot. Pulled out of deep poverty by her own determination and her intensely competitive character, she became one of the best aviators in the world. As she put it, "I might have been born in a hovel, but I determined to travel with the wind and the stars."[2]

Before she was able to do that, though, she had to put in a great deal of time on an earth that was definitely unwelcoming. Jacqueline Cochran never knew exactly when she had been born nor who her parents were. She grew up with foster parents in Columbus, Georgia, and her dresses were made out of old flour sacks. When Jacqueline was eight, her family moved into a house with the first bathroom she had ever seen. That same year, she wore her first pair of shoes and also took her first job: the twelve-hour night shift in a cotton mill. Even at that age she was determined to manage her life better than her foster parents had managed theirs, and she soon moved up to a better job in a beauty parlor. Cochran tried nursing but soon decided that as a nurse she could never make enough money to help herself or anyone else. She returned to the beauty business, and moved to New York City to sell cosmetics.

Cochran might never have become interested in flight if she hadn't seen it as a useful aid in launching her cosmetics business. The idea itself might never have occurred to her at all but for a chance remark at a party by the man she would later marry, millionaire Floyd Odlum. He told her, half-jokingly, that if she really wanted to succeed in the business she would need wings. To Odlum's surprise, Cochran acted on his suggestion. She visited various cosmetics companies and asked them, "If I learn to fly at my own expense and go on a country-wide tour to advertise your products, will you help support the expense of my plane?" Enough agreed to this odd idea that Cochran took a vaca-

[2] Cochran, Jacqueline, *The Stars at Noon*, p. 46.

tion at the Roosevelt Field Airport, earning her pilot's license after only two-and-a-half weeks. Armed with the license, she took off for San Diego and learned the equivalent of the Navy's flight-training course.

Cochran would later say that when she had paid for her first flying lesson, "a beauty operator ceased to exist and an aviator was born." Actually, it was her business that flourished, not her flying career, for a few years after she had received her license. It was not until July, 1937, that she really saw success as a pilot.

In 1937 Cochran spent most of her time beating women's speed records. In July she set a new women's national record; in September, a new women's world record. In December she set a new national transcontinental record—beating Howard Hughes's earlier one—by racing from New York City to Miami in only four hours and twelve minutes.

The New York-Miami flight was a dangerous and frightening one in many ways. The plane Cochran flew was designed to hold two extra fuel tanks: one was behind the seat, and the other was the seat itself. But the plane had never been tested before, and no one had discovered that the tank behind the seat was off the center of gravity. Because of this the plane's nose could not be kept level. It pointed either straight up or straight down until the offending tank had run dry. Cochran had planned that flight down to the last detail. Not wanting to weigh the plane down, she had measured its fuel so thriftily that the last tank ran out the exact instant that she landed in Miami.

Taking such risks was second nature to Cochran. Her husband said of her that although she ran wildly at the sight of snakes and became hysterical listening to ghost stories, she had never feared real danger; instead she generally moved instinctively to the center of trouble. It was not surprising that in 1937 the International League of Aviators voted Cochran the world's outstanding pilot, an award she would win for three consecutive years. She also won the Harmon Trophy in both 1938 and 1939 and continued to break record after record for years after World War II. (In 1962, on a flight from New Orleans to Bonn, West Germany, she set forty-nine speed records on the same day.) No matter how grueling any flight was, she always tried to comb her hair and put on makeup before leaving the plane. This very human act annoyed reporters and other aviators, but it must have made her seem as if nothing could fluster her, not even the most dangerous flights.

One cannot mention Jacqueline Cochran's success as a record-setter without returning to the story of Jacqueline Auriol. The two women have constantly been compared, and while both are irritated by the comparison, it is certainly true that they have spent a great deal of time snatching records back and forth from each other. Between 1951 and 1963, Auriol claimed the women's world speed record five times; five times Cochran took it back. And the two women were both so ambitious that the success of one undoubtedly spurred the other to work harder. "Once the competition had started," Auriol explained in her autobiography, "we had to keep going."

Auriol's first world record was the one made almost two years after her accident. In 1951 she flew a jet, a Mistral, at 818 kilometers (about 515 miles) an hour, beating Cochran's previous record. Auriol's flight, made so soon after such a grave accident, was crucial in establishing her reputation as a serious pilot. But Cochran did not accept the flight, since it had been made in a jet. In a statement to the New York *Times*, she explained, "The splendid French flying woman did not beat my record, which I obtained with a plane of a different type . . . The record obtained with a jet propelled plane cannot annul the record obtained by a plane with a motor explosion type of engine." Cochran's statement sounds polite but determined. Her protest had no effect, however, and Auriol's time was officially accepted. Jacqueline Auriol was now the fastest woman in the world.

The year after her first victory, Auriol improved her record time, flying in a Mistral at 855.920 kilometers per hour. The year after that, Cochran recaptured the record in a jet, beating Auriol by almost 200 kilometers per hour. The battle was on.

In 1955 Auriol set a new world record, and there was an interesting story behind it. On May 20, she received an invitation to a celebration in New York City: Pan American Airways, of which Cochran's husband Floyd Odlum was a principle shareholder, was sponsoring eight days of festivities in honor of a record number of transatlantic passengers. Auriol had already accepted the invitation when she learned, completely by chance, that the Fédération Aéronautique Internationale, whose vice-president was Jacqueline Cochran, was about to abolish women's records as being demeaning to women. Cochran herself held the women's world record for speed. When the law went into effect, she would be the last woman to have held the world record; in effect it would become hers permanently. Cochran also knew that France owned a Mystère IV N, a plane that could beat her own Sabre. The new Fédération law was to take effect in eight days. If Auriol went to New York for the Pan Am celebration, she would be at parties during those eight days. She decided instead to stay at home and train, and to recapture the women's record before she lost the chance for good.

Fate must have been on her side, for everything about the flight seemed to fall into place: Auriol was quickly given permission to fly the Mystère, and a team of twenty people was placed at her disposal. Auriol made only one mistake during the flight itself. She lost her bearings for a second and announced "I'm lost" to ground control. This naturally sounded as though she were in mortal danger, and it frightened a good many people. When she landed, Auriol was told first that she had beaten Cochran's record and second that her careless use of language during the flight meant she would not be permitted to fly for one month. Nevertheless, she had won a victory for herself and France.

Jacqueline Cochran saw to it that Auriol was awarded the Harmon Trophy for this flight. The following year, Cochran became president of the Fédération Aéronautique Internationale. One of her first acts as president was to reinstate women's records. She then proceeded to beat Auriol's record in 1961, flying at 1262 kilometers per hour.

Auriol was now forty-three years old, and she knew she would not be making record flights much longer. She was hugely tempted to try once more and achieve a speed so great that the record would remain with France for a good long time. She decided to fly the Mirage III, in which she had already achieved Mach II—twice the speed of sound. She assembled a team of twenty-three specialists whom she called the best of the best.

To train for the flight, Auriol rose at dawn and made one or two flights in the morning before eating lunch; then she rested before flying at dusk. She made seven trial flights, six of which were record-breakers. It was clear at once, she said, that she could beat Cochran with her eyes closed. Indeed her official time, 1849 kilometers per hour, beat not only Cochran's record but also that of Gerard Muselli, who held the men's record. And Auriol decided that, after all, she had not finished setting records. After a full year of training, she went on to beat her own record by two hundred kilometers.

There is always the chance, for the women who set records, that a new plane, begging to be tried, will be developed; or that a rival will set a new record too tempting not to beat; or simply that the aviator herself will find she does not want to stop. Jacqueline Cochran once wrote, "I have had my fling in the jet phase of aviation. The rocket is just around the corner. Will Father Time let me wait for this? I hope so."[3] It is in this spirit that so many women have devoted themselves to setting records, and have changed the face of aviation.

[3] Ibid., p. 236.

British Amy Johnson flew her Gypsy Moth solo from London to Port Darwin, Australia in 1930. The nineteen-day trip was filled with problems; at one point Johnson was forced to repair her plane's wings with shirts borrowed from a nearby school. Her ground engineer's license (the first given to a woman) enabled Johnson to make all repairs alone.

Frances Marsalis set a record in aviation and in human patience when she spent almost ten days—including her birthday—aloft in 1933.

Amelia Earhart, by far the world's best-known woman pilot, poses in front of a map charting a proposed flight.

When women aviators began to achieve real fame, it was inevitable that accounts of their escapades reached the world of fiction. This dramatic still is from the 1933 film Christopher Strong, in which Katharine Hepburn played the lead. The film was loosely based on the biography of Amy Johnson.

Hélène Boucher was one of France's most important woman aviators in the 1930s. In August 1934 she set three speed records one after the other, finally setting a record which made her the fastest pilot in the world. Only a few months later she was killed in a plane crash.

Jacqueline Cochran rose from extreme poverty to become one of the greatest aviators of either sex. For three years in a row she was named the world's outstanding pilot; she was crucial to the American war effort during World War II; and she set records of all kinds for years after the war.

*J*acqueline Auriol, the "other Jacqueline," was considered Jacqueline Cochran's greatest rival by the public. Auriol is shown here alighting from the Mirage in which she flew at twice the speed of sound and beat both the men's and women's world speed records in 1961.

*T*his steely-eyed woman is Laura Ingalls, who once threatened to shoot a reporter who ventured too close to her plane.

F*rance's Adrienne Bolland, the world's first pilot to fly over the Andes, is shown here after setting another world record: she looped the loop 98 times in less than an hour. It was not dizziness but a leak in the plane's fuel tank that finally forced her down.*

B*obby Trout prepares her plane for an endurance flight. Endurance flying, one of the most demanding forms of record-setting, was mainly popular among women aviators before World War II.*

B*eautiful Jenny Dare was the first woman aviator to have a comic strip all to herself. Jenny had a great deal to keep her busy. She flew in races and pylon contests, generally beating her male opponents, and made rescue missions and other dangerous flights with her friend and sometime copilot Wanda.*

*E*linor Smith, the "Flying Flapper of Freeport," after a flight. Smith soloed when she was only fifteen and managed to set both endurance and altitude records. High-altitude flights of the day required that the pilot be dressed as warmly as possible, even though this made it difficult to move.

The Races & the Racers

THE FIRST NATIONAL air race for women took place in 1929. Nineteen women took part in the race, gathering together in their planes on August 18 at Santa Monica. As they waited for the starting signal, some of America's greatest women pilots were treated to the privilege of listening to Will Rogers making jokes about them for the benefit of the press. One of the contestants, Amelia Earhart, recalled that the press took its cues from Will Rogers and had finished making up funny names for the derby and its pilots before the racers had reached their first stop. The pilots were rechristened "Angels" and "Sweethearts of the Air," and the first Women's Air Derby turned into the "Powder Puff Derby."

At least the Sweethearts of the Air had the consolation of knowing how well they had performed in the Air Derby. The race was organized by the National Exchange Clubs, who had originally planned to allow any woman to enter and to bring a mechanic with her. Soon the organizers realized that a few too many Hollywood starlets seemed to be entering. None of them could fly, but their mechanics could. Obviously Hollywood was trying to provide itself with some publicity: the mechanics would do all the work of piloting while the actresses took the glory. At this point the National Exchange Clubs decided to change the eligibility rules. Any woman entering would need a current license and 100 hours of solo flying.

Probably only thirty women in the country fulfilled those requirements in 1929. The fact that twenty of them entered the race indicates how eager women aviators were to be taken seriously. Only seven American women had transport licenses; six of them flew in the Air Derby. Sixteen of the twenty entrants finished the race. There had never been a cross-country derby with such a high proportion of racers who completed the course.

The aviators in the race could also be pleased that they had been able to convert so many people to the idea that women could fly. Many of the onlookers who gathered to watch at various points along the race were women who had never before seen other women in airplanes. Some of these women were so eager to learn about flying that they jabbed their umbrellas through the fabric of a few planes' wings to see inside.

This first Women's Air Derby was part of the National Air Races based in Cleveland. Most of the contestants were American, though Thea Rasche of Germany and Jessie Keith-Miller of England also raced. The race was scheduled over eight days, with prearranged refueling and overnight stops.

Many of the racers, tired out after long days of flying, were distressed to find that they were expected to appear at long dinners in their honor when they stopped for the night. As Will Rogers said, "They've had to land in every buffalo wallow that had a Chamber of Commerce and put up a hot dog sandwich." Usually the meal that actually faced them was the standard banquet chicken. Some of the women began to send cables ahead with the succinct message: "No chicken, please."

Californian Margaret Perry was forced to leave the race early when she came down with typhoid fever. Three other women did not complete the derby. One of these was Florence Barnes, who accidentally smashed her plane into a car near a runway. This must have been especially disappointing to Barnes, because she had already had to find her way back onto the course after getting lost in Mexico. Ruth Nichols also had to drop out of the race after hitting a tractor. Nichols had had an excellent chance of doing well in the race, and she won the admiration of the other racers by refusing to complain about her misfortune.

The third woman who did not finish the race was Marvel Crosson. She was first noticed missing at the stopping point in Phoenix. After a short search, Crosson's plane was found crashed in the desert. A few feet away lay her body, trapped and tangled in her parachute. Crosson had evidently tried to jump from the plane when it was too close to the ground to permit the parachute to open.

Marvel Crosson's death nearly ended the Air Derby. One newspaper headline about the accident read, "Women Have Conclusively Proven They Cannot Fly." Many people tried to call off the race, but the racers were adamant about continuing. Amelia Earhart spoke for all of them when she said, "Marvel Crosson left a challenge to the women of the Derby and there is certainly no aftermath of fear among us." Fortunately the race committee was equally determined that

the race should go on. The committee released the statement that its members wished "officially to thumb our noses
at the press," and the race continued.

A series of minor misfortunes dogged the remaining
racers. Most of them were handled with remarkable aplomb.
Ruth Elder, for example, had to make an emergency landing in a pasture. Elder had chosen to fly in the Air Derby
in an effort to improve her reputation with her fellow pilots.
Earlier in her career, she had tried to fly the Atlantic only
days after receiving her pilot's license. She had not succeeded, and some aviators thought Elder had only tried the
flight in order to win herself some quick fame. She hoped
that flying in the Derby would show the world that she
was a serious pilot.

Elder's flight was certainly a challenge. She was flying
along without any trouble when a sudden gust of wind tore
her map out of her hands, "leaving me clutching a piece
about as big as a postage stamp." Elder finally decided to
land and ask for directions. Below her she could see a
pasture that looked all right for a landing. She did not
notice until she was on the ground that she was surrounded
by cattle, all of them very interested in her bright red plane.
"I prayed," she told the other racers later. "I said, 'Oh God,
let them all be cows.'" Elder left the pasture ungored and
came in fifth in the heavy-plane division.

Fourth place among the heavy-plane pilots went to
Blanche Noyes from Ohio, who had had her share of trouble
en route as well. While flying above the Texas desert, Noyes
noticed smoke curling into the cockpit. Directly behind her
was a fire in the baggage compartment. She could not reach
her fire extinguisher and had to make a crash landing and
put out the fire with handfuls of sand. Noyes kept her composure until she reached the airfield for the scheduled stop
at Pecos, Texas. Her plane looked, as Louise Thaden said,
"like a wounded duck with a broken wing and badly crippled

legs." Noyes emerged silently from the plane, her face black with smoke, and held up her scorched hands. When she was asked how she had managed to get the plane off the sand and into the air, she began to cry. "I don't know," she said.

Amelia Earhart also had an accident while landing at Yuma, and it damaged her propeller. A new propeller was flown in in time for her to finish third in the heavy-plane division. Second place went to Gladys O'Donnell from California, who had had only forty-six hours' flying experience in her life.

During the first leg of the race, Louise Thaden was almost knocked out by carbon monoxide poisoning from her airplane's exhaust pipe. She had to spend the rest of the flight bending forward to breathe from a four-inch tube feeding oxygen into the cockpit. It was an uncomfortable position to fly in, but maybe it brought her luck: Thaden won the Air Derby easily, arriving in Cleveland after thirty-one hours and nineteen minutes in the air. Her winner's speech was an unusually gracious one. "I'm glad to be here. All the girls flew a splendid race, much better than I. Each one deserves first place, because each one is a winner. Mine is a faster ship. Thank you."

The 1929 Women's Air Derby made the women who flew in it realize how much they could benefit from organizing themselves. When the race was over, four pilots— Margery Brown, Fay Gillis, Frances Harrell, and Neva Paris—sat down together and wrote a letter that they sent to all licensed women pilots in the United States. On November 2, 1929, twenty-six women gathered at the Curtiss Airport to make plans for forming the first all-woman aviation organization.

After several suggestions for what to name the new group (including the Homing Pigeons, the Climbing Vines, and the Noisy Birdwomen), Amelia Earhart stood up and proposed that the organization be named for the number

of charter members who joined it. This idea was immediately accepted; Earhart was named president of the new
group; and when the new members had all been counted,
the name of the organization became the Ninety-Nines.
(There are now 164 chapters of Ninety-Niners throughout
the world, with more than five thousand members in all.)

When the Women's Air Derby was held in the following
year, Gladys O'Donnell came in first in the heavy-plane
division. (Phyllis Fairgrave Omlie won for light planes, as
she had in 1929.) O'Donnell flew from Long Beach, California to Chicago, a distance of 2,245 miles, and won
$3,500—a huge sum in the Depression. She then decided to
stay in Chicago and enter the National Air Races, in which
she won first place in four of the women's events. One of
the events she won was a women's fifty-mile free-for-all
with a prize of $2,500. About this race the New York
Times wrote admiringly, "With the exception of Mrs.
O'Donnell, none of the seven contestants was able to
make the banks and turns that indicate good aimship."[1] In
the same year, O'Donnell also won second place in the
Tom Thumb Derby. On January 27, 1931, the National
Aeronautical Society presented her with the Aerol Trophy,
one of the highest honors in women's aviation.

In England, the 1930 King's Cup race, the country's
most celebrated cross-country derby, was won by a woman.
Winifred Brown flew an Avro Avian for 750 miles and beat
eighty-seven other competitors. Brown was the first woman
ever to win the race, and her victory had a special savor for
two reasons. The first reason was that Brown had flown
with a copilot who was there to fly for her if she became
tired, and she found that she did not need his help after all.
The second was that Brown had been in an accident in 1928
that had caused a boy's death. Though the accident had not

[1] Adams, Jean, and Kimball, Margaret, p. 262.

been her fault at all, Brown had considered leaving aviation for good; she had only continued so that it wouldn't appear that she was in fact responsible for the death. The King's Cup race was a major event in convincing the public that Brown had been right to keep flying.

In 1931, for the first time, the National Air Races opened the same events to both men and women. The races were staged by the city of Cleveland. They included a cross-country derby from Santa Monica as well as many events in Cleveland. The racers flew over the same cross-country course and stopped at the same places, but men and women were awarded separate prizes. However, one special prize was to be awarded to the pilot of either sex who amassed the greatest overall number of points in the races.

For the derby, the pilots' planes were handicapped according to the top speeds they had reached. Gladys O'Donnell led both men and women for the first leg of the derby, and she also led four days later on August 24. Although she was pleased to be ahead, O'Donnell was worried as well. She couldn't believe that she was beating the men in the race fairly, and she told the racing officials (adding her voice to those of many men flying in the race) that the handicaps must have been computed wrong.

Fortunately the judges—mostly men—assured O'Donnell that she was wrong. One said firmly, "Handicapping has nothing to do with the situation. The girls are flying better —that's all." This was far from the only dispute in the 1931 races, however, and not all of them were as easily settled. Every night of the derby was highlighted by a protest meeting held by most of the eighty-one racers. T. H. Kinkade, the official handling all the protests, had a nervous breakdown when the racers reached Kansas City.

After the races were finished, the pilot with the highest number of points overall was Phoebe Fairgrave Omlie. She won a total of $12,000 in prize money—the most awarded to any of the racers—and won a car as well.

Mae Haizlip won second place in the 1931 derby, and she also won second place in four of the races. Haizlip won a first in an additional race. At the 1932 National Air Races, Haizlip also beat Ruth Nichols's world speed record, flying in a plane in which her husband had recently set a transcontinental speed record. One of the Haizlips' sons sitting in the grandstand remarked that his mother's new record—an average of 252 miles per hour—was "all right, but Dad can go faster."[2]

Mae Haizlip's triumph at the 1932 races was one of the few for women that year. The women aviators were pitted against the best men in the country in eight free-for-all events. Most of the men were flying specially-designed racing planes. This fact, added to the fact that there was no handicapping, meant that the women in the race had virtually no chance of winning any of the male-female events. Moreover, the press seemed less favorably disposed toward women pilots in 1932. The aviation editor of the *Cleveland Press* wrote, "Watching the race, it was easy to tell why women do not wish to compete with men pilots on an equal basis."[3] The next year's National Air Races offered only two events for women. Despite the fact that two new women's races had been established that year—the Amelia Earhart Trophy Race and the Shell Speed Dashes —the National Air Races were still the most exciting to the nation, and the 1933 races were a harsh blow to women pilots.

In 1934, officials of the National Air Races announced that no women at all were to be allowed to take part.

[2] Ibid., p. 263.
[3] Planck, Charles E., *Women with Wings*, p. 89.

Furious but undaunted, the Ninety-Nines organized their own 1934 meet at Dayton. They called it the Women's National Air Meet, and it attracted much more publicity than the men's races that year.

This was most gratifying, but the Women's National Air Meet also suffered tragedy when one of the pilots was killed. Francis Marsalis was turning around a pylon in a pylon race (a race around a course marked by towers called pylons) when her plane crashed and she was killed instantly. The future of women's racing looked grim for several months after the accident, and despite the overall success of the Women's National Air Meet, officials doubted that women would be allowed back into the National Air Races. Finally it was decided that women could once again take part in the men's races, largely because of their excellent performances in other races that year—and probably also because the male pilots feared being upstaged again.

The following year saw an emphatic turnaround for women racers. The United States Bendix Trophy Race, one of the most important transcontinental races in the world, had recently begun to allow women to participate, with a special $2,500 prize for the first woman to complete the course regardless of her place in the race itself. This was a thoughtful but rather patronizing gesture; it implied that no woman could possibly actually win the race. But on September 4, 1936 a woman did win the Bendix—Louise Thaden.

Thaden and her copilot Blanche Noyes planned to take off from New York City at 4:30 A.M. But they found themselves unable to sleep, and after a brief discussion about who should jump out of the plane first if they had an accident, they decided to leave earlier. As they climbed into their plane, Thaden's husband told her that he thought she had an excellent chance of coming in second or third as long as she didn't get lost.

Several other women were preparing to take off at the same time. One was Laura Ingalls, the only woman pilot that year to fly the Bendix solo. Ingalls made it a point to be completely independent during the flight, fixing everything on the plane herself and refusing all offers of help. In fact, she refused to let anyone come near the plane. When an over-inquisitive reporter tried to interview her, Ingalls was terse. When he walked too close to the plane to suit her, she whipped out a pistol and snapped that she would use it if he came any closer. Ingalls had the satisfaction of finishing second in the race, and she could say she had done it all herself.

Thaden and Noyes did not get lost, but they did run into a storm near St. Louis. The race's only requirement for qualification was that every contestant must arrive in Los Angeles by 6:00 in the evening, Pacific Standard Time. Both Thaden and Noyes were worried that they would not meet the deadline. To keep up their spirits, they made a bet. Thaden predicted that they would cross the finish line at 5:11, Noyes predicted 5:08.

Their plane was flying at 230 miles per hour as Thaden and Noyes approached the Los Angeles Airport, and the plane swept across the airport so fast that Thaden had no time to look at the ground. She jumped as Noyes suddenly thumped her in the arm: the plane was speeding away in the wrong direction from the finish line. Thaden jerked the plane around so quickly that both women went faint.

"Dodging Marine Corps planes, we crossed the white finish line at right angles to the grandstands, triumphantly —from the wrong direction, but we crossed it!"[4]

It was 5:10, so the pilots called off the bet. After crossing the finish line, they began to fly as inconspicuously as they could down to the end of the field. They didn't

[4] Thaden, Louise, *High, Wide and Frightened*, p. 180.

think anyone would be interested in them; both were convinced that they were among the last arrivals. Before they had gotten very far, though, the plane was being followed by a few men, and then it was surrounded by a huge crowd, all shouting and jumping up and down. Thaden and Noyes landed to see what was the matter. One man ran forward and opened the door of the plane: "Get out of there, we think you've won the Bendix!"

They had won it. They had flown from New York to Los Angeles in fourteen hours, fifty-five minutes, and one second, winning the race and setting a new transcontinental speed record at the same time. The name of the $2,500 prize for the first woman to cross the finish line had to be changed. Before the race, it had been called the "Consolation Prize." Now it became known as a "Special Award."

After the Bendix had been conquered by a woman, women's air racing finally achieved real status in the eyes of the people who had laughed at it before. One of the many benefits this conferred on women racers was the fact that it became much easier for them to obtain the superb planes men had always been able to fly in races. Jacqueline Cochran won the Bendix in 1938, having flown in a Seversky Pursuit nonstop from Burbank, California to Cleveland in eight hours and ten minutes. Her success, and the success of the plane, made it possible for Seversky to develop the pursuit planes that were subsequently used in the second World War.

When the war began, civilian flying had to end for the duration. Interest in women's aviation dropped tremendously after the war and remained low for several years. To try to revive that interest, a new series of Powder Puff Derbies (by this time the nickname seemed permanent) was started in the United States in 1947. The derby was officially called the All-Women Transcontinental Air Race, or AWTAR. Many women promised to appear in the first race in 1947, but on the appointed day a single plane was the only one on the starting field. Its pilots, Caroline West and Beatrice Medes, flew the course anyway and came away with the trophy.

Fortunately there were more planes in the 1948 race—six of them. Frances Nolde, the vice-president of the National Aeronautical Society, won the race. And in 1949 things looked far more hopeful. Seventeen planes flew in the AWTAR, and almost a hundred women participated in the 1949 All-Women Air Show.

There was great diversity among the pilots who raced after the war. Both amateur and professional pilots raced, and there was also a great age range among the pilots. Two sixteen-year-olds—Deborah Diemand and Becky Greer—flew in the AWTAR, each taking her mother along as co-pilot on the flight. Another woman named Betty Gillies flew in the AWTAR several times in the 1950s with her teenage daughter Pat as co-pilot.

In 1960 Viola Gentry raced in the AWTAR. Forty-two years had passed since she had set the first women's solo endurance record. Women's air racing had begun to become healthy, and this process still continues.

On August 21–25, 1979, a race with one of the longest names in history took place: The Angel Derby's First Women's Air Derby Fiftieth Anniversary Commemorative Race. Once again the racers flew from Santa Monica to Cleveland. On hand to receive a commemorative plaque was Blanche Noyes. She was the only person who made any jokes about the race, and this time nobody minded.

Pylon racing is one area in which there are still very few women. The pylon races in the United States are held annually in two places: Mojave, California, and Reno, Nevada. The Reno races are generally far better-attended than the ones in Mohave, because Reno itself is such a tourist attraction. (In fact, there have been a few years in which the Mohave races have had to be cancelled due to lack of funds.)

The only two American women who now fly in pylon races are Judy Wagner and Colene Giglio. Giglio has a full-time career in aviation: she runs her own airline company in California. Wagner's career is in medicine: she assists her husband as an oral surgeon. Wagner's plane was built by her husband, and in it she won the first pylon race ever won by a woman, in October, 1979.

There is another series of races that is definitely out of the aviation mainstream. Betty Pfister, a distinguished American airplane and helicopter pilot, also enjoys travelling in balloons. She is part owner of the Columbine, a helium balloon named after the state flower of Colorado. In 1977 Pfister set up the first hot air balloon race, and balloon races have been held annually since then, with more than thirty balloons going aloft for the event.

CHAPTER 8

"I Have Freed a Man to Fight"

T HE SECOND WORLD WAR was the first war to make
effective use of aviation. It was also the first war
in which women played a crucial role. These
dubious achievements had at least one positive
result, which was that aviation in general—and women's
aviation in particular—made great advances because of the
war.

It took some time, though, for women in the Allied
countries to convince their governments that they should
be included in the war effort. Most women's suggestions to
that effect met either with outright disapproval or with
patronizing laughter and a pat on the head, just as they do
now. "Man, how the women like to play soldier!" com-

mented one writer in 1942, and his response was pretty standard.

Many women pilots in the Allied countries were eager to be put to use in active combat, but their hopes were rarely fulfilled. The British government was willing to use women aviators wherever it could, however, and British women were involved in the war far earlier than were their American counterparts. Though few people actually liked the idea of using women pilots to help fight the war, no one could deny that doing so was practical: It helped free male pilots for active combat.

Britain went a step further than the United States would when it granted its women aviators full military status. (American Helen Richey, who had gone to England to lend her piloting skills, became a major before she was discharged.) Women gained military status only six months after the war began. Soon after the British Air Transport Auxiliary was formed, military officials realized that here was one way women pilots could be helpful. A women's division of the ATA was formed, and Pauline Gower, who had been flying for over ten years, was picked to command it.

When Gower assumed command, there were eight women besides herself in her division. When the war ended there were more than 100—one-fourth of the entire ATA—from all over the world: Great Britain, the United States, Canada, Poland, Norway and even Turkey. Of course the women had been nicknamed by the press. They were called the "Ata-girls," a title they hated—but no one criticized their performances.

Actually, the Ata-girls were very popular with the public. A British newspaper wrote about them in these approving words: "Women and men ferry pilots work on terms of absolute equality. There was some idle talk about the glamorous wealthy girls, all owning their own planes, who had

rushed to join the Air Transport Auxiliary at the outbreak
of the war, but were inclined to be more ornamental than
useful. Nothing could be further from the truth. Several
of them are married to wealthy men, or belong to the
Social Register, but for sheer hard work and endurance they
do their share and more."[1]

The women in the Air Transport Auxiliary ferried new
military planes (Hurricanes, Lancasters, and Mosquitos)
from factories to eagerly waiting air bases. This service
sounded distinctly unglamorous, but it was tremendously
useful. Women ferry pilots flew the planes to where they
were desperately needed; they also "rescued" damaged
planes from anywhere the planes had been landed and flew
them to repair areas. The latter task was by far the more
dangerous one. The women flying the damaged planes
could not always tell exactly how much damage had been
done to the planes until they took off in them, and they
repeatedly risked serious accidents.

While of course many women came to the Air Transport
Auxiliary with piloting backgrounds, not all of them did.
Rosemary Rees, who delivered over 200 four-engine bombers
during her stint with the ATA, had been a ballet dancer
before signing up. Mona Friedlander was a professional ice
hockey player before she became an Ata-girl. And the ATA
provided many women with flying experience, which proved
invaluable to them later, both in and out of the military.
Margot Gore, relatively inexperienced when she joined the
ATA, became Commanding Officer at the Hamble ferrying
division. Joan Hughes became the only British woman during the war who was qualified to give flying instructions in
all kinds of planes—even the largest bombers.

One of the most famous women in the Air Transport
Auxiliary was Amy Johnson, who had tried to join without

[1] Planck, Charles E., Women with Wings, p. 261.

any publicity and who tried to remain inconspicuous once she was a member. Her service to England ended tragically on freezing, stormy January 5, 1941. Johnson was flying over the Thames when the plane she was ferrying broke down. She parachuted into the icy Thames and began to struggle towards a patrol ship, *H. M. S. Hazlemere.*

The patrol crew tried to bring the *Hazlemere* closer, but the fierce winds stranded the ship; the crew then threw ropes to Johnson, but they landed just out of her reach. As the men began to lower a lifeboat, the *Hazlemere's* captain, Lieutenant-Commander Fletcher, dived into the river and swam towards Amy Johnson.

Fletcher reached her before her heavy, waterlogged flying clothes pulled her under. He had hoped to be able to hold her up in the water until the lifeboat reached them, but the water was so choppy that the little boat began to fill with water almost immediately. Johnson was now too weak to move, and when she slipped out of Fletcher's exhausted grip he could not retrieve her. Both the captain and Amy Johnson disappeared under the waves before the lifeboat ever got to them.

Fifteen women in all died while working in the Air Transport Auxiliary, but this did not deter the rest. In fact, their safety record was better than that of the men in the ATA; Pauline Gower had seen to it that the Ata-girls understood that—as she put it—they were paid to be safe, not brave.

In 1944 the ferrying service began to take planes across the English Channel. Diana Barnato became the first woman in the ATA to ferry a plane to France. She flew alongside her husband, and each of them delivered a Spitfire. This set off many other trans-Channel deliveries for women. Ata-girls also brought damaged aircraft back into England for repairs. In addition, they were among the country's first women to pilot jets, which England began to use shortly before the end of the war.

Other nations in the British Commonwealth had their own women's aviation groups. Canada created the Canadian Women's Royal Air Force in July 1941, though few members of this group actually flew. Australia established the Women's Auxiliary Australian Airforce in the same year. Only a few months after its creation, the organization had fifty officers and almost 700 members. By 1945 there were approximately 600 officers and 18,000 members.

It was time to establish a women's auxiliary to the Air Force in the United States. At least women thought so. Verna Burke, a flight instructor, expressed the feelings of most women pilots in the country when she said that the United States should train women "to ferry ships, trainers, bombers, ambulance ships—all kinds, and relieve the men who are more fitted for combat flying. But train them now, as soon as possible. The women in England have proven they can stand up under long hours of bombing and hard work. Why not give the American women who are really sincere about flying a chance to prove that they can do as well?"[2]

Few men in the military agreed. In 1941, Jacqueline Cochran returned to the United States from England, where she had been training some American women to work in the Air Transport Auxiliary. Cochran was fired up with the idea of starting a similar group in the United States, but it was Nancy Harkness Love who first came up with what the Army considered an acceptable idea for getting women aviators involved in the war.

Love's plan probably worked because she was not over-idealistic. She knew that women pilots had virtually no chance of assuming great power within the American military, much less of entering actual combat. Even to suggest such a thing would seem too threatening to the Army.

[2] Ibid., p. 262.

Instead, Love worked on the idea that women could be helpful: not important, not powerful, but simply useful.

Nancy Love pointed out that there were enough American women who could fly airplanes of at least 200 horsepower, and who had at least 500 flying hours to their credit, that it might be practical to put them to work for the Ferrying Division. More pilots were needed to ferry planes to England; why couldn't women fill in the men's vacant spots and do ferrying in the United States? Such women would not even need to be considered part of the Air Force. They could remain civilians and be paid by civil service.

It seems hard to believe today that such an eminently practical idea caused any trouble at all. Eventually the United States Air Force (or "Airforce" as it was then spelled) agreed to try the plan, and Nancy Harkness Love became the first woman to fly for the Air Force.

Love may have succeeded partly because her husband was deputy chief of staff of the Air Transport Command. But her own credentials were stellar as well. She had learned to fly at the age of sixteen. By 1942, she had amassed one thousand five hundred flying hours, some of them grueling ones. In 1937 and 1938 she had been a test pilot for Gwinn Aircar Company. Among other things, she had had to test the landing gear of new aircraft. This task required her to slam planes onto the ground as hard as she could, in order to see if they would hold up. In addition, she was able to fly regularly, twice a day: she commuted to and from work in her Fairchild 24.

In 1940 Nancy Love began ferrying planes herself, flying them to the Canadian border where Canadian pilots picked them up and flew them to Europe. It was then that she began pushing her idea about letting women help the Air Force. In 1942, the Women's Auxiliary Ferry Squadron (WAFS) was created with Love as head.

On September 6, 1942, Love sent a telegram to 200

American women pilots, informing them that the Ferrying Division Air Transport Command was looking for women members: "Advise Commanding Officer Second Ferrying Division of Air Transport Command New Castle County Airport Wilmington Delaware if you are immediately available and can report at once at Wilmington at your own expense."[3]

The first trainee to arrive at Wilmington was Betty Huyler Gillies, who came the day after she had been sent her telegram. Gillies had already made a name for herself in aviation. In 1929 she had helped to found the Ninety-Nines, in 1933 she had been named vice-president of Gillies Aviation Corporation, and in 1940 she had convinced the Civil Aeronautics Association that pregnant women should not be denied the right to fly.

But Betty Gillies was far from being the only skilled pilot among the WAFS trainees. Over 100 women applied to the WAFS after receiving Love's telegram, and only twenty-seven were picked to be the first trainees—the "Originals."

Women who were picked for the WAFS had to pass a flying test and an Army physical examination. It was also necessary for them to prove that they would function well as team members; Nancy Love was not interested in any prima donnas. Once accepted, WAFs learned a huge variety of things: Air Force flight procedures, meteorology, navigation, codes, ferrying responsibilities and—not least important to a fledgling organization trying to ingratiate itself to the military—military law and military courtesy. WAFS trained physically as well, and they even learned how to march in a formal parade. As commanding officer of the whole process, Nancy Love was a great favorite with the trainees. In general she proved excellent at the job. She had picked women who

[3] Crane, Mardo, "The Women with Silver Wings," *The 99 News*, p. 9.

could function well as a team, and she took care to be the same way herself. Though she expected every one of the WAFS to work as hard as possible, she never asked too much of them, and she never made one of "her girls" fly a ferrying mission that she herself would not have been capable of making.

There was one area in which Nancy Love admitted she was faulty. The fault was a rather endearing one: she found herself completely unable to shout orders at the WAFS. During drills in which she led the formations, she some-times became so self-conscious that she forgot what to say. One day, when the WAFS were drilling on an empty run-way, she realized that they were heading right toward a ten-foot drop at the runway's end. Love became so flustered that she couldn't give a single command. She watched tongue-tied as all twenty-four women in the formation— "roaring with laughter," she recalled—marched to the em-bankment, jumped down, and kept marching, leaving her standing alone at the top, "still speechless!"[4]

By October 22, the "Originals" were ready for their first ferrying assignment, delivering Cubs (training planes) to various Air Force bases. Having proved that they could do this without mishaps, the Originals moved on to ferrying heavier planes while new trainees continued to ferry training planes. Nancy Love became the first member of the WAFS to ferry a bomber when she and her copilot, Barbara Towne, flew a B-25 from California to Cincinnati. After that, mem-bers of the Originals ferried more bombers and pursuit planes. Barbara Erickson made four two-thousand-mile ferry-ing flights in just five days, an accomplishment for which she won the Air Force Medal.

It was clear to all except the staunchest traditionalists that women in the WAFS were performing a necessary job

[4] Ibid., p. 12.

most skillfully. They were the only women's service group at the time that was exposed to the same dangers as men's groups; they underwent exactly the same training as men pilots who ferried for the Air Transport Command. But because the WAFs were civilians, they were given no government compensation in case of accidents.

When one of the Originals, Evelyn Sharp, was killed during a failed takeoff in a fighter plane, there was no money to pay for her funeral. Worse, there was no insurance money available to take care of Sharp's invalid mother, who had been completely dependent on her daughter for support.

It was not so in every country that took part in World War II. Russia's women pilots were not only given full military status, but they were also given combat posts. In fact, they were the single group of women pilots most often to come under fire during the war.

For about ten years prior to the war, the U.S.S.R. offered intensive aviation training to a group of women who had joined a civil defense organization called the Osoaviakhim. Women in the group learned and practiced flying, gliding, and parachuting in their spare time; most of them had regular occupations in other fields. When war broke out, the training of the Osoaviakhim proved invaluable.

Once Russia was at war, Stalin asked Marina Raskova, one of the most celebrated women aviators of the day, to form and head the 586th Fighter Wing and two other women's fighting units as well. The 586th was a wing of women fighter pilots whose main job was to guard Russian industrial areas. The second women's unit consisted of bombers and the third of dive bombers. Marina Raskova had already achieved world fame by setting an international long-distance nonstop flight record with Valentine Grizodubova and Polina Osipenko. Under Raskova's command, the three women's wings were crucially important to the Soviet war effort.

Lilia Litvyak was one of the most famous pilots in the 586th Fighter Wing. Twenty-one years old at the time she joined the unit, Litvyak was so small that she needed pillows on her plane's seat just so that she could see out of the cockpit. Though she was diminutive, Litvyak was a tremendously effective fighter, and she fought with all the more determination because she had had to watch as Germans gunned down her fiancé's plane.

On her third patrol flight, Litvyak shot down two German planes. She was to bring down several more before her death in August, 1943, on the Soviet-German front. Litvyak was killed in action when her plane was shot down, but not before she had managed to bring down seven Nazi planes.

Two sisters in the Soviet Air Force also made important contributions to the war effort. Militsa Kazarinova flew an attack plane and was chosen to be chief of staff of her own women's bomber regiment. Her sister, Lieutenant-Colonel Tamara Kazarinova, commanded her own regiment and flew as a fighter pilot as well. For this work she was awarded three decorations, one of them the Order of Lenin.

Flyers Tamara Pamyatnykh and Taisa Surnachevskaya functioned perfectly as a fighter team. Faced once with a formation of forty-two Nazi bombers, the two women instantly flew straight into the formation and began to shoot. In minutes they had shot down four Nazi planes, and they continued to shoot. The formation broke up. The Nazi planes scrambled to speed off as quickly as possible, jettisoning their bombloads as they raced away, and the two women, though wounded, managed to land safely.

Any list of Russian women's piloting accomplishments in World War II seems to consist solely of remarkable numbers. Marina Chechneva, a combat pilot, made 810 combat flights; Natalya Mekin, a member of a women's light bombing regiment, had 980 combat flights on her record. Major of the Guards E. Nikulina dropped 150 tons of bombs over the course of the war; Guards Lieutenant Nina Lobkovskaya was decorated for bravery four times. Clearly, women aviators could take on combat roles without any trouble.

Germany's Hanna Reitsch had an exceptional role as a wartime test pilot for the Nazis. Reitsch was later to be remembered for flying the last airplane out of Berlin before Germany's collapse in 1945, but her other accomplishments were just as memorable.

Hanna Reitsch's entire career was one of the most spectacular of this century. It began pretty unspectacularly, though, and it began secretly: Like many other beginning pilots, Reitsch felt unable to tell her parents how much she wanted to fly.

Reitsch was born on March 29, 1912, in Silesia. Her early ambition was to become both a pilot and a doctor (her father was a doctor), and in her late teens she enrolled secretly at the gliding school in Grunau. (Gliding assumed more importance in Germany than in many other countries because the Treaty of Versailles at the end of World War I had limited the use of powered aircraft in Germany.) On her very first day there, she plunged her glider to the ground nose down because she had not followed instructions closely enough. The incident was humiliating, especially because Reitsch overheard the school's director say that the school had better dismiss her before she killed herself.

Reitsch was ordered grounded for several days. She crept miserably to bed that night and lay awake worrying. Suddenly it occurred to her that she could practice gliding right there in the room: all she needed was a walking stick. She got up and found one, climbed back into bed, and sat there pretending to be in the pilot's seat of a glider. Using the walking stick as a control column, Reitsch began to make a practice flight. "In imagination, I now slid forward, keeping my direction and balancing so that neither wing dipped towards the ground. Then the plane came to a stop. I knew that I had carried out a faultless ground slide."[5]

After a second and third night of practicing with the walking stick—during which she imagined herself "making little hops" and thirty-foot ascents into the air—Reitsch had gained remarkable new confidence in herself. When she was again allowed to take the control column of an actual glider, she astounded her classmates and the school's director.

Her progress from then on was rapid, but she had some trouble convincing her parents, once she had confided in them, that they should let her keep flying. Finally they agreed that she could continue gliding only if she also attended medical school. She dutifully agreed and began studying medicine in 1932, the same year she obtained her flying permit. As it turned out, though, she studied medicine as little as possible. (In her first term she read none of the assigned books.) By 1934 she had come to realize that all she wanted to do was fly. The idea of medicine faded away; instead Reitsch became a research-and-test pilot at the Darmstadt Air Research Center for Gliding.

On a sunny May day in 1933 Reitsch went up for a trial flight of the Grunau-Baby, the newest training glider model. She decided to try to fly blind, using only the glider's instruments, and she also decided not to bother putting on a coat because the day was so warm. She was so absorbed in the instruments that she did not notice a huge storm cloud until she was directly under it—almost three thousand feet in the air. But she was not dismayed. Rather, the sight of the cloud intrigued her. She had always wanted to fly through one.

Reitsch climbed almost two thousand more feet and then, suddenly, "a million drumsticks suddenly descended

[5] Reitsch, Hanna, *Flying Is My Life*, p. 12.

on the glider's wings and started up, in frenzied staccato, an ear-splitting, hellish tattoo."[6] The cloud had begun to storm. Ice and hail, added to the extreme height of the glider—9,750 feet—froze the instruments Reitsch had been relying on so heavily. She realized that she was no longer the glider's pilot but merely its passenger. She would have to rely on its "inherent stability" to get her safely to earth.

Reitsch did land safely, partly because of that stability and partly because she did take hold of the control column when the glider started to fly upside down. She had set a new unofficial altitude record and had come to know during the flight that whatever she went through for aviation was worth it.

The flight had other rewards as well. After setting the record, Reitsch was offered the chance to fly as a stunt pilot in a film. During her spare time on location she set two new unofficial glider endurance records for women; she also earned money to help finance her 1934 scientific gliding expedition to Brazil, where she studied the South American winds and set several women's altitude, endurance, and distance records.

Hanna Reitsch's career was constantly punctuated with firsts. She had already been the first woman to glide upside down; in 1937 she became the first glider pilot of either sex to cross the Alps. The same year she became Germany's first "Flugkapitän" (flight captain) for the German commercial Air Service.

Her first important war activity came when she tested new types of freight-carrying gliders. Reitsch tested the first transport glider ever developed at the Glider Research Institute. Its first load was a sack of sand; Reitsch then added another sack each flight until she felt secure enough to substitute a human passenger for the sand bags.

[6] Ibid., p. 46.

In the 1943 film Flight for
Freedom, *Rosalind Russell*
played a daring aviator
much resembling Amelia
Earhart. (*The film's heroine*
crashed her plane on
purpose to avoid Japanese
capture.) *RKO Studios*
never stated outright that
the film was based on
Earhart's life, but they did
buy permission to make the
film from Earhart's
husband.

A class of WASPs poses at Avenger Field, Sweetwater, Texas in 1944. This was a select bunch: less than four percent of WASP applicants were actually chosen.

Nancy Harkness Love was the first woman to fly for the American Air Force in World War II. She was also the first to come up with a workable plan for using women pilots in the American war effort. Love was chosen by the Air Force to direct the Women's Auxiliary Ferry Squadron (WAFS), whose main function was ferrying war planes to Air Force Bases. American women aviators were not allowed to come near real combat.

Jacqueline Cochran lolls against the propeller of a plane during World War II. "A strong coequal Air Force" was indeed Cochran's main goal during the war years. She headed the Women's Airforce Service Pilots (WASPs), a group of dedicated aviators who ferried planes and officers, towed targets, learned to fly bomber planes, and tested pursuit planes. The WASPs were finally given veteran status in 1977.

In World War II women were more likely to decorate airplanes than to fly them.

Germany's Hanna Reitsch was one of that country's greatest pilots before, during, and after World War II. She began by flying a glider and was the first glider pilot, male or female, to cross the Alps; she was the first Flight Captain for the German Commercial Air Force; she test-flew the ME 163 rocket and the V-1 flying bomb during the war; and she flew the last airplane out of Berlin before Germany fell in 1945.

Soviet women aviators were far more active in World War II than were their counterparts in other Allied countries. These two sisters, Militsa (left) and Tamara Kazarinova, flew for the Soviet Airforce for twenty years. Both were pilots in the war, Militsa an attack-plane pilot and Tamara a fighter-pilot; each was also head of a women's regiment.

From left to right: Polina Osipenko, Valentine Grizodubova, and Marina Raskova. These three women became world-famous when they flew nonstop for 3,671 miles, an international record. Raskova also became immensely helpful to the Soviet Union in World War II when she formed and headed three women's fighting units.

Not until the 1970s did American women become fully accepted in the military service. These four women were among the first to be trained as Navy pilots. From left: Ensign Rosemary Conatser, Ensign Jane Skiles, Lieutenant JG Barbara Ann Allen, and Lieutenant JG Judith Ann Neuffer. Neuffer was the first woman to pilot a plane through the eye of a hurricane.

Anesia Pinheiro Macahdo was Brazil's first woman pilot, having begun flying in 1922, and she kept an active pilot's license for more than 50 years—another women's record.

Geraldyn, or Jerrie, Cobb *was an important figure in American aviation of the 1950s. She became a professional pilot at the age of 18. After holding a great many kinds of piloting jobs—including crop dusting—and setting many speed and altitude records, she became in 1960 the first American woman to take and pass NASA's tests for becoming an astronaut.*

In *1978, NASA chose 35 astronaut candidates for its new space shuttle program. Six of them were women, the first American women ever chosen as candidates. Here they are shown being briefed during training at a survival school in Oklahoma. From left: Margaret Rhea Seddon, Kathryn Sullivan, Anna Fisher* (whose husband is also an astronaut candidate), *Sally Ride, Shannon Lucid* (obscured), *and Judith Resnik.*

During a training exercise, Sally Ride practices wearing a parachute harness and ejection gear and learning what it feels like to be suspended in midair.

Here Marie Stivers helps revive old-fashioned wing-walking, an ordeal she compares to "using a hurricane for a hair dryer." (The pilot's name can best be read when the plane is upside down in the air.)

Blanche Stuart Scott, the first woman to make a solo flight, was honored more than seventy years later by a commemorative airmail stamp issued by the United States Postal Service.

Next Reitsch tested a new glider brake system. The Germans planned to invade France in February, 1940, and they needed to be able to bring gliders to a quick stop when they landed on ice. Reitsch wrapped herself in blankets when she tested the first such brake system, but even so it proved more powerful than she had planned for: when she landed, she was snapped against her safety belt so brutally that her wind was knocked out of her. Such inconveniences did not bother her, though. "The test flights which I carried out for the Glider Operation, as, indeed, my whole work at this time, had as object the saving of human lives and for that reason alone, quite apart from my love of flying, I could not have wished for a more satisfying task."[7]

Reitsch was also instrumental in German experiments in balloon-cable cutting. To protect itself from Nazi bombers, the British military had created the Balloon Command, sending up hundreds of huge barrage balloons, which prevented bombers from approaching their targets. Of course, Reitsch saw the Balloon Command from a different perspective and felt that the barrage balloons were an "unforeseen obstacle" to be removed before they harmed any more German planes (balloon cables often cut right through planes' wings). Accordingly, she was pleased to be able to test planes specially equipped to deal with barrage balloons.

Fenders on the front of these planes protected them; scissor-like devices on the wingtips cut the balloon cables. Though she caught scarlet fever during the test period, Reitsch continued to test the new equipment as long as she could. In doing so she lowered her resistance, and the fever spread to her eyes. Shortly after that she developed muscular rheumatism as well. The antibarrage-balloon testing had to be taken up by another pilot, but Reitsch re-

[7] Ibid., pp. 176–177.

turned to it as soon as she was able. She received the Iron Cross, Second Class, for these tests and so became only the second woman ever to receive that honor.

Reitsch then went on to a new kind of flying entirely. In 1942 she began to test-fly the rocket-powered Messerschmitt ME 163. On her first flight, she reported, "It was all I could do to hold on as the machine rocked under a ceaseless succession of explosions. I felt as if I were in the grip of some savage power ascended from the Nether Pit. It seemed incredible that Man could control it."[8]

On only her fifth test flight in the ME 163, the rocket crashed. Moving slowly and with huge caution, Reitsch drew a little sketch to show how the crash had occurred. Next she tied a handkerchief around her face to spare her rescuers the terrible sight it presented. Only then did she lose consciousness. This time Reitsch was awarded the Iron Cross, First Class.

Reitsch was also the only woman to fly the V-1, the flying bomb introduced in the last year of the war as one of Hitler's "vengeance weapons." This creation was a small jet missile carrying explosives and guided by a gyroscope. Only Reitsch's extremely small size permitted her to fly one, which she did in an effort to learn how the V-1 performances could be improved.

Jacqueline Cochran had by no means left the scene in the United States. In September, 1942, she was named director of a new pilots' group at a salary of a dollar a year. As director, Cochran would be responsible for training the women to ferry aircraft for the WAFS. The name of the new pilots' group, which trained at a base in Sweetwater, Texas, was the Women's Flying Training Detachment. Ten months later the U.S. War Department declared that all women pilots in the military, trainees and ferry pilots alike,

[8] Ibid., p. 190.

were to be placed under Cochran's command. (Nancy Love continued directing the women in the Air Transport Command.) Moreover, they were all to be given a new name—Women's Airforce Service Pilots, or WASPs.

By 1944, twenty-five thousand women had sent in applications to become WASPs. Less than four percent of these women—1,830—were selected, and they learned to be ready for anything. They ferried planes and ferried officers and towed targets so that airmen could improve their aim. WASPs also flew as test pilots. It was a WASP, Ann Baumgartner, who flew the first jet used in the Air Force, and many WASPs were called in to test planes male pilots had been afraid to fly.

Seventeen WASPs were selected in 1943 to learn how to fly the B-17, or "Flying Fortress"—a huge four-engine airplane and the most important World War II bomber. After arriving at Lockburne Air Base, the seventeen women looked at each other and realized that Cochran must have chosen them for their size. Some were six feet tall; the group as a whole averaged five feet eight inches. This may have made their male instructors feel more confident about teaching women, although the few extra inches could hardly have made a difference in an airplane weighing twenty tons.

WASPs had to learn how to fly the Flying Fortress as if they had been born knowing how. On each flight they were given what their instructor called "blindfold cockpit checks," when they had to be able to point at any instrument in the plane with their eyes closed and had to say whether the plane was climbing or descending by listening to the engine. The WASPs even made "hooded" takeoffs. On these their instructors covered the plane's windscreen with a black curtain, forcing the women to rely only on their instruments.

The WASPs learning to fly B-17s were also tested in other

areas. They were the first women to fly above twenty-five thousand feet in unpressurized planes when the Army began to wonder what happened to women pilots at high altitudes. At thirty thousand feet one WASP's fountain pen exploded in her pocket. Another woman's watch exploded (she had forgotten to take it off before climbing), but the women themselves were all fine. The Army gave each a certificate qualifying her to fly above thirty-five thousand feet.

Not every WASP experience was an unalloyed success. In July, 1943, twenty-five WASPs reported for a top-secret mission at Camp Davis, a fifty-thousand-man Air Force base in North Carolina. The WASPs joined 600 male pilots to learn how to tow targets for gunnery trainees. These targets, made of twenty-foot pieces of muslin, were attached to the planes' wings. Pilots flew target planes back and forth along the sunny North Carolina beaches as gunners took turns shooting at their targets.

The assignment turned out to be one of the most grueling the twenty-five women had ever faced. Camp Davis mechanics told the WASPs, in confidence, that only three planes on the base were really trustworthy. The women at first thought this was a joke, but they realized gradually, with horror, that it was true. It was almost impossible for the base to receive repair parts for damaged planes, and in addition the base had been allocated such a small amount of fuel that most of the planes flew on fuel with a lower octane level than they really required. (Some WASPs suspected that sabotage efforts had been made on their planes as well. Again and again they found water in the carburetors of the A-24s they flew.)

The women at Camp Davis were understandably terrified of flying planes in such bad shape, especially on night missions—and especially because some of the gunners were so inexperienced. Some felt betrayed and angry as well as frightened, certain that Jacqueline Cochran, interested only in using them as test subjects, had abandoned them.

On August 23, 1943, WASP Mabel Rawlinson burned to death at Camp Davis when her A-24 cracked into two pieces in the air and burst into flame. Mechanics discovered that the A-24 had had a faulty canopy latch that had trapped Rawlinson inside the cockpit.

After Mabel Rawlinson's death, the WASPs stationed at Camp Davis told Cochran about everything wrong with the target-towing operation there. They found her a rather unsympathetic listener. Determined to continue the mission—a setback might ruin the whole women's program—Cochran did not act on their protests, and two WASPs resigned as a result.

Twenty-three new WASPs, unaware of what faced them, were sent to Camp Davis five days after Rawlinson's death. Only one month passed before the base saw a second tragedy when Betty Taylor was killed in a crash. This time Cochran investigated the accident more thoroughly. She kept quiet about what she had found, sure that to publicize the findings would result in women's being sent away from the Air Force immediately. Cochran had discovered that the gas tank of Taylor's A-24 had contained sugar, which could cause an engine to break down almost instantly.

Had Betty Taylor died because of sabotage? If so, Cochran did not want Taylor's fellow WASPs to find out and panic. The death had already dealt enough of a blow to the WASP program at Camp Davis. All WASPs flying there in large planes were thereafter required to take copilots.

Much more successful was the 1943 program in which WASPs tested and ferried pursuit planes. Ten women were chosen for this, preparing at the Air Transport Command pursuit school in Palm Springs for several weeks before taking off in P-47s and P-51 Mustangs—the largest and fastest pursuit planes in the nation. The first class of WASPs finished training in January, 1944, and a steady stream of WASPs was sent to learn how to fly pursuit planes after that. It was an impressive step for women's aviation: those WASPs helped introduce the United States to a new kind of flying. The new pursuit planes went so fast that many of the first women to fly them accidentally overshot their destinations by hundreds of miles. But that wasn't much to worry about. The WASPs' accident rate in these planes was much lower than that of their male counterparts.

WASPs were sometimes used by the Army to improve male morale and to spur male pilots into giving better performances. Dora Dougherty and Dorothea Johnson were once given three days of secret, highly intensive training in order to learn how to pilot the B-29 Superfortress. The plane was aptly named, for it was the most expensive, most intricate, and most massive aircraft that had ever been made. Dougherty and Johnson flew the B-29s so well that the Chief of Air Staff in Washington ordered the two women to stop flying it at once. They were putting men to shame, he said. Though they were never again allowed to touch the controls of a Superfortress, Dougherty and Johnson had the consolation of knowing that they had helped to win acceptance for the Allies' most impressive war plane.

> When the General comes, Sir, to view us in our drill
> We'll do a four winds march, Sir, and check out o'er
> the hill,
> And when he calls "Attention!" we'll click our heels
> and yell,
> "I'm just a raw civilian, Sir, and you can go to hell!"
>
> —Thelma P. Bryan

It was a terrible blow to the WASPs, and perhaps even more to Jacqueline Cochran, when Congress turned down a 1944 bill that would have granted military status to the women who had helped the Air Force so much. Once the bill had been turned down, the end of the WASPs came quickly. The Air Force decided to end the WASP training program as well, so the program was deactivated on December 20, 1944. It was deactivated though WASPs had volunteered to continue working for one dollar a year; though the Ferry Command had begged to be allowed to continue using its women pilots, who now ferried seventy percent of all single- and twin-engine fighter planes; and though the invasion of Normandy had taken place only days earlier.

At a farewell ceremony in honor of the WASPs, Jacqueline Cochran said, "The emotions of happiness and sorrow are pretty close together, and today I am experiencing them both at the same time, as well as the third emotion of pride. . . . My greatest accomplishment in aviation has been the small part I have played in helping make possible the results you have shown."[9] A few days later WASPs turned in their uniforms and watched sadly as Army officers' wives scrambled through the discarded uniforms to see what they could take for themselves.

But the WASPs were never entirely forgotten. In 1977, Senator Barry Goldwater introduced a bill "to provide recognition to the Women's Airforce Service Pilots for their service to their country during World War II." This recognition, it was hoped, would come in the form of making WASPs official veterans of the war. After sharp opposition, Congress finally passed a revised version of the bill. The WASPs were assigned veterans' status in November 1977, and soon afterwards they were issued official honorable discharges.

[9] Van Wagenen Keil, Sally, *Those Wonderful Women in their Flying Machines*, pp. 301–302.

Military Women
After World War II

T
HOUGH AMERICAN WOMEN had clearly been useful
in many capacities during the Second World War,
their acceptance into the military in the ensuing
years could only be described as sluggish. Some
important advances took place within five years after the
war, but they were followed by a long period of relative
inactivity. The first step in giving women pilots entrance
into the armed forces came three years after the war ended,
in 1948. This was the passage of the Women's Armed
Services Integration Act, which ensured women of both
regular and reserve status in the Army, Navy, Marine Corps,
and Air Force.

However, the act went only so far: It restricted the

number of women in the military to only two percent and limited the rank of women officers as well. One woman in each armed service component could become colonel director or captain director of the component, but that was the only time women were permitted to hold such a high rank. In all other cases women could not achieve ranks higher than lieutenant colonel or lieutenant commander.

After the Women's Armed Services Integration Act had passed, Air Force women organized a group called, appropriately enough, Women in the Air Force, or WAF. Geraldine May was the colonel director of this group. May had been a WAC—a member of the Women's Army Corps—during World War II and had served with the War Department general staff after that.

May was fortunate in being able to head an organization that could not be tossed out as the WASPs had been. The WAF was not an auxiliary to the Air Force; it was a vital part of the Air Force itself, and so were its members. But the power of the WAF was limited severely by the fact that its members were not permitted to become navigators, bombardiers, or pilots.

February, 1949, saw the Women's Auxiliary Air Force in England replaced by the Women's Royal Air Force. The new organization, called WRAF for short, trained right alongside the men in the Royal Air Force, and members of the two groups worked together as well.

Women in the WRAF had a distinguished Commandant-in-Chief: none other than Queen Elizabeth. Their Air Chief Commandant, hardly less distinguished, was the Duchess of Gloucester. Felicity Hanbury was the first Director of the WRAF. Hanbury had been a private pilot and had held several jobs in the old Women's Auxiliary Air Force. One of her first acts as Director of the WRAF was to visit the United States to take a look at the Women in the Air Force. She urged American women to try for a less limited role in the Air Force; England had proved to be much less conservative than the United States about this matter.

Hanbury's advice had little effect for a long time. Women took active part in the armed forces of many countries, but it was not until the 1970s that American women began to do the same. In 1967, the law limiting the percentage of women in the armed forces was repealed. Women could now be promoted to flag rank, could serve in the National Guard, and could receive promotions and other benefits at a rate roughly comparable to that of men.

In 1966, Ensign Gale Ann Gordon became the first woman to solo in a Navy training plane. In 1973, the Navy became the first service to open aviation training to women when it picked eight women for its flight training program. Four of these women were civilians who had just graduated from college; four were already officers. In all, six of the women completed training.

Those six were under close scrutiny for their entire training period. The Navy offered no more flight training for women until it could ascertain how well the first class had done. Eight more women were admitted in 1975, and again, six completed the program and earned their wings.

The third class of women to begin flight training did so under different circumstances. They reported directly to the Naval Aviation Schools Command at Pensacola, Florida, where they were given sixteen weeks of aviation officer candidate training. This was the first time that the school had ever taught women.

Rosemary Bryant Conatser, one of the women in the Navy's flight training class (and one of the very few military pilots to fly a Powder Puff Derby), summed up the opinions of many of her classmates this way: "Why do I want to go to a tactical squadron, to fly off a boat, and perhaps be shot at? My reasons are the same as those that have always attracted men to Naval Air. It is because I have experienced the satisfaction of the first step—winning Gold Wings—and I want to continue to succeed at what is the most demanding form of aviation. . . . The sense of joy is as much with me now as on my first solo. It has taken different forms as I mature. What was once a quest for fun has developed into the desire to be a professional naval officer, learning to handle responsibility, with command in mind."[1]

Lieutenant Judy Neuffer was one of Conatser's classmates in flight training. Neuffer came from a family with a strong aviation tradition: Her father had been a World War II Army Air Corps pilot, and he had given her her first flying lessons when she was fifteen years old. Neuffer was the first woman ever to fly a plane through the eye of a hurricane. When asked how it had felt, she admitted candidly that she had been much too busy flying the plane to notice anything about her feelings. She was also candid about one of the disadvantages of working in such a male-dominated field: "Sometimes I really long to have a woman to go to lunch with or to talk woman things with. I just can't go to the wardroom and sit down and talk about woman things with the guys."[2]

The Army was next to open aviation training to women. By 1979 there were forty women Army airplane pilots and approximately fifty women helicopter pilots. Training wasn't easy for any of them. Helicopter pilot Lieutenant Karen Anderson was the only woman in flight school at Fort Rucker, Alabama. One of her classmates, who was supposed to be her copilot, told her he would never fly with her: "First of all, you're a woman and I don't think you should

[1] Collins, Helen I., "From Plane Captains to Pilots," *Naval Aviation News* (July, 1977), pp. 17–18.
[2] Nye, Sandy, "Up Front with Judy," *Naval Aviation News* (July, 1977), p. 20.

be in the Army. Secondly, I don't think you should be flying."[3] Anderson found another copilot. (When a field grade officer ordered her to dance with him, Anderson told him he would have to put the order into writing.)

Army women had to learn to be just as tough as Army men. When Karen Anderson had problems making her male subordinates salute her, she stood in front of them and made them practice saluting over and over until she was satisfied. It generally took about fifteen minutes before she was satisfied, she said.

Janna Lambine was the only woman at the Coast Guard Officer Candidate School in Virginia. She was also the first woman to become a pilot for the Coast Guard, having applied for flight training while she was in officers' school. Lambine was designated a Naval Aviator on March 4, 1977. To date only two other women have become Coast Guard aviators. Like Janna Lambine, Lieutenant JG Colleen Annee Cain pilots a Coast Guard helicopter, flying search and rescue flights and pollution and fish patrols. Lieutenant Vivien Suzanne Crea pilots a Hercules C-130 at Barbers Point Air Station in Hawaii.

In 1976 the Air Force opened its Academy to women for the first time and began its first training program for women pilots. This was a significant move, and there was a lot of flap about it. The women fit in very well, however, with a minimum of fuss. They knew what the rest of the Air Force soon found out—that they would be able to serve as pilots as well as their classmates.

Twenty women in undergraduate pilot training were picked for an Air Force test program "designed to identify any training deficiencies or operational problems peculiar to women in pilot training." Very few deficiencies were found.

[3] Smith, Linell, "In Control," The Baltimore *Evening Sun* (August 15, 1979).

"The women are going through exactly the same training as the men and are hacking it well,"[4] said Lieutenant Colonel C. T. Davis of Williams Air Force Base.

Pilot trainees in the Air Force, both women and men, face a great deal of work. The training program lasts for almost a year. Each student undergoes several weeks of ground training before being allowed to fly. After nineteen flights with an instructor pilot, trainees are finally allowed to solo in T-37s. They then spend four months learning everything there is to know about T-37s before they move to six months of instruction in T-38 Talons. Captain Connie Engel was the first Air Force woman to solo in a T-37; First Lieutenant Christine Schott, one of whose classmates said she had the true fighter pilot's personality, was the first woman to solo in a T-38 in May, 1977. When asked about the difference between the two planes, Second Lieutenant Carol Scherer said, "The T-38 has the power when you need it. The T-37 goes, but when you want more power you have to wait for it. The T-38 *really* goes—now!"[5]

To most women training to become Air Force pilots, the hardest part of training is not the flying time but the time spent in classes. Captain Kathy La Sauce said, "It's all the bookwork that really drags you down."[6] Students must spend 350 classroom hours learning, among other things, aerospace physiology, accident prevention, navigation, applied aerodynamics, and weather. Their days generally begin at four-thirty in the morning and do not end until late in the afternoon. There is little free time in such a schedule.

Kathleen Cosand was another woman to begin Air Force

---

[4] Arnold, Maj. Terry A., "Baptizing the New Breed," *Airman* (October, 1977), p. 5.

[5] Ibid., p. 6.

[6] Ibid., p. 5.

pilot's training in 1976. As an undergraduate, Cosand, a member of the Reserve Officers Training Corps, fully intended to become a pilot despite discouragement both from her classmates and from the ROTC itself. The ROTC at first refused even to let Cosand take the pilot's qualifying test, but she protested constantly until they gave in. She was vindicated when she received perfect scores in the sections of the test devoted to piloting and navigation.

Passing the qualifying test was only one of the Air Force requirements Cosand fulfulled without trouble. More worrisome to her was the fact that she was not quite tall enough to fit the height requirement. When the time came to be measured, Cosand stretched as hard as she could and succeeded in making herself the required five feet, four inches tall.

In addition to learning how to fly T-37s and T-38s during her training period, Cosand spent nine weeks training to fly a C-141, a jet transport plane. The C-141 was so much larger than the airplanes she had flown previously that flying in it made her airsick for some time.

Like many other pilots in training, Cosand's private life virtually disappeared. Two days after she was married to a fellow pilot, the couple was sent on separate Air Force assignments. After their first two months of marriage they had managed to see each other for only one week and had made hundreds of dollars worth of phone calls to each other. Nor could they wear their wedding rings. Pilots are forbidden to wear rings when they fly.

By the end of the decade, the sacrifices demanded by serving in the Air Force were slowly beginning to seem worth making. Over 100 women held command positions, six women as brigadier generals. Air Force women worked as pilots, navigators, air-traffic controllers, and Air Force security police. Still, less than ten percent of the Air Force personnel were women, and it was estimated that by the mid 1980s only one in eight Air Force jobs would be held by a woman. Moreover, it is still most difficult to predict when women in all branches of the military service will be assigned to combat positions.

The question of whether or not women should be placed in combat is probably the most controversial area in the history of women's aviation. As early as the First World War, women aviators asked to be allowed to fight and were turned away. The issue is more complicated now. Though most people are reluctant to face the thought of women at war, the fact is that American women are presently assigned to jobs that could place them under fire in the event of an attack. There are now more women in combat-related jobs in the United States than in any other country. What are the chances of their actually taking part in combat?

Both Army and Air Force officials think it is very likely. An official Army document states: "Women are not peace-time soldiers. Women may be required to engage and to be engaged by the enemy, experience physical contact with the enemy and experience a high risk of capture." The Army considers its women to be eligible for ninety-five percent of all Army jobs, and in 1977 it was officially announced that Army women would now be permitted "to accomplish unit missions throughout the battlefield." Recently the Army also announced that it would allow women to serve in the elite 82nd Airborne Division, a paratrooper division in North Carolina, and to other units that had formerly been closed to them and that could easily expose them to attack.

The Army has found—as have the other services—that one of the main problems with women in combat is one of logistics. It is much more difficult to plan living arrangements for two sexes than it is for one. (One Army officer found as well that women in his unit gave markedly poorer performances if they were forced to go without showers for more than three days.) More serious is the fact that many women are not strong enough to operate heavy machinery. To date this problem has not been solved.

The Air Force nonetheless plans to use more women in the near future than any of the other services. It currently bars women from only four of its 230 job categories. "We expect that our women will experience combat; it's a fact of life," said the director of Air Force personnel planning. To that end, some Air Force women have begun flying cargo transports for the Military Airlift Command and tankers for the Strategic Air Command. In addition, the Strategic Air Command has recently chosen women for training with crews handling the Titan intercontinental ballistic missile, which would be launched in the event of a nuclear attack.

None of these preparations ensure exactly what the role of women aviators in combat will be, nor do they clarify what that role should be. While being permitted to serve in combat positions may represent a political achievement for women, it does not represent an equally obvious moral step forward. Perhaps women who want complete equality with men should be required to take part in combat; on the other hand, perhaps the aggression needed to perform well in battle is not such a virtue that it should be instilled equally in every member of society. Whatever happens in the future, it is certain that women in aviation and elsewhere are beginning to know the responsibilities as well as the rewards of improving their status.

Postwar Civilian Flight

THE SITUATION for professional women pilots has changed a great deal since Helen Richey became a copilot for Central Airlines in 1934. Eight men applied for the job as well, but Richey—fresh from helping to set a world endurance record and winning the 1934 Women's National Air Meet—was the one chosen. This was a first for women in aviation. No woman had ever flown as a pilot on regularly scheduled flights, and for a long time after Richey's brief period on the job, no woman did so again.

Richey realized soon after beginning work for Central Airlines that things were not going to go smoothly for her. Male pilots hated having her around and said so. The pilots' union refused to consider her for membership. Soon the airline itself seemed to change its mind about employing a woman and announced that none of its women pilots would be permitted to make flights in bad weather. Of course the airline had only one woman pilot, and she had no doubt of what the company intended. Richey stayed with Central Airlines for a little while longer and then resigned before they could humiliate her further.

Perhaps made cautious by Helen Richey's experience, Marilyn Grover Heard of Kansas announced in 1948 that she was looking for work as a free-lance copilot. This meant that she could fly for any airline interested in hiring her without being tied down to one in particular. In turn, the airlines using her services would not feel threatened by having a woman pilot, because she would only be a "temporary." After Heard had flown on several free-lance jobs, her strategy—if that's what it was—worked, for in 1949 Transocean Air Lines hired her as a regular copilot.

There are now forty-five thousand American pilots in the airline industry. Fewer than 200 of them are women—in fact, it is estimated that there are no more than 150 women airline pilots in the Western Hemisphere—but the number is growing. And at least the women who fly professionally today have the security of knowing that they will be able to keep their jobs.

It was not until the 1970s that the airlines began to open their doors to women pilots, and then only gradually. Emily Warner, probably the third woman airline pilot in the United States, spent seven years trying to be hired by a commercial airline. During that period she worked as a flight instructor and applied over and over to the airlines, who rejected her over and over in return. She was finally hired by Frontier Airlines as a first officer in 1973 and currently works as first officer on a Boeing 737.

Flight captain Ernie Clark was unaware when he took

his daughter Julie for rides that she would decide to follow in his footsteps—or wings—one day. But in 1977 Julie Clark Ames became a copilot with Hughes Airwest, an airline her father had cofounded. In doing so she reached a goal for which she had been striving for many years; she had always wanted to work for Hughes "because of the heritage behind it all." Ames did this even though her father had died in the country's first hijacking while piloting a plane thirteen years earlier: the hijacker burst into the cockpit and shot him. The plane crashed and killed everybody aboard.

Ames was still determined to fly after her father's death. Her first step was to leave college in her senior year and go immediately to Trans-World Airlines, where she was chosen for training as a flight attendant. TWA did not know that Ames had fibbed about her age to be accepted for training. When the airline found out, only two days before graduation, that Ames was six months younger than she had claimed to be, it was too late for them to do anything about it. Ames had already been trained, and besides, she was due to receive a prize for best trainee.

The activities leading up to her job at Hughes were varied. In 1969 she decided it was time to learn how to fly. She spent her summers working as a flight attendant and her winters working on her ratings. (To earn the money to pay for some of these ratings, Ames imported antique Japanese clocks—bought from a man she had met while serving as a flight attendant in Korea—and sold them to Navy families.) After receiving her commercial certificate as well, Ames worked for a time air-delivering tractor parts to farmers in California and Nevada. She took time out to fly the 1976 Powder Puff Derby with Dianne Mann as her copilot. Next came a job as a flight instructor for a Cessna dealership. Ames piloted charter flights and sometimes, to her dismay, was asked by her boss to serve drinks to the passengers as well.

Though the job with Cessna was not ideal, Ames was ultimately grateful for it. It led to her being hired by Golden West Airlines, the largest commuter airline, as their first woman pilot. Ames did not settle down after being hired but instead went on to earn her jet certificate. She still hoped to work for Hughes, but there was one obstacle— her husband was also a Hughes pilot. This difficulty was neatly resolved when the airline's only other woman pilot decided to marry a Hughes pilot. Julie Ames was hired soon after that.

In working to become a Hughes pilot, Ames followed a route typical of many women who fly today. She worked for her long-term goal in stages, remembering her ideal but also remaining practical. She took what jobs were available, no matter how far-removed they were from what she wanted. Each new job gradually took her further away from the traditional women's role in aviation and brought her closer to the untraditional job she finally found. That is, untraditional from society's point of view; but after all, Ames was continuing what her father had done before her. Every time she flies now, she wears his wings inside her coat.

Another woman whose father helped inspire her choice of career is Jean Haley, a United Airlines pilot and president of the International Social Affiliation of Women Airline Pilots. Haley's father was a crop duster who did not live to see his daughter become a commercial pilot; he was killed while dusting a field, but she continued to fly.

At present the youngest woman airline pilot in the United States is Kim Goodman. Goodman soloed at the age of sixteen and received her private pilot's license on her seventeenth birthday. On her eighteenth birthday she received her commercial license, on her nineteenth her helicopter rating. She was hired by Western Airlines as their thirteenth woman pilot; Goodman was twenty-two at the time.

American women pilots have things comparatively easy. There are presently only two women pilots among the one thousand five hundred male airline pilots in Mexico: Conchita Bernard and Elena Folch. Bernard once said ruefully: "Everywhere we go they always call us sweetie."[1] Though grateful for their jobs, the two women have come to dread the strain of the accompanying publicity. They are called almost every day for interviews. The situation is no better for France's Anne-Marie Peltier, who flies for AirInter and is one of only six women airline pilots in that country. Peltier feels the problem might be one of too little publicity. In an interview with the Los Angeles *Times* (May 16, 1980), she said, "People in France don't know about us women pilots at all. They don't know we're up there."[2]

Women are also holding more jobs than ever as flight instructors. One of the earliest to begin doing so after the war was Charlotte Kelley, who originally received a degree in education from the Massachusetts School of Art during World War II. After graduation she decided to get a job as a flight attendant instead. On her days off she learned how to fly, becoming such a skilled pilot and navigator that she began to teach at Boston's Aviation Training School in 1949.

In subsequent years, Kelley learned to pilot helicopters and seaplanes. (When she heard that she had passed her seaplane pilot's test, she managed to thank her instructor before falling off the plane's pontoon into the water.) Kelley also flew the biggest blimp of the 1950s: the Navy's ZPG-2.

In 1955 Kelley became the first American woman ever to fly a National Guard jet. The jet's pilot, who had piloted

[1] Hillinger, Charles, "Women Airline Pilots Share Achievements," The Los Angeles *Times* (May 16, 1980).
[2] Ibid.

the plane that dropped the atomic bomb on Nagasaki in World War II, handed Kelley the controls when the jet was at twenty-seven thousand feet, and she flew for about a quarter of an hour. She also became the first American woman to cross the sound barrier as a passenger, flying in an Air Force Starfire in 1957. (She reported that all she felt while breaking the sound barrier was a small "buffeting" sensation.)

Brazil's first woman pilot also held the oldest active pilot's license of any woman in the world. Anesia Pinheiro "Shorty" Macahdo learned how to fly in 1922 and continued for over fifty years. She received both Brazilian and American pilot's certificates; she worked as a flight instructor for both commercial airline pilots and military pilots. In 1972, the Women's International Association of Aeronautics awarded her the Lady Hay Drummond-Hay–Jessie R. Chamberlin Memorial Trophy for her exceptional career.

THE WHIRLEY-GIRLS first got together in 1955. They were aviators of a different sort: They flew helicopters. Helicopters were not invented until World War II, and the first person to demonstrate one to the public was Hanna Reitsch.

Reitsch's introduction to the helicopter had come in 1938. A year earlier, Germany's Professor Focke had invented the first helicopter and had created a sensation. There had never before been seen an aircraft that could hover, fly both backwards and forwards, take off vertically from the ground and land in exactly the same spot.

Many people around the world could not believe that such a machine existed, and so the Germans decided to hold public indoor helicopter exhibitions in Berlin to prove that it did. For three weeks Reitsch demonstrated the wonders of the new machine. During the three weeks, both the helicopter and Reitsch had received tremendous publicity.

Ann Shaw was the first American woman to be licensed as a helicopter pilot, receiving her certification in 1947. Shaw also became the world's first woman to fly a commercial helicopter; she worked as a pilot for the Metropolitan Aviation Company, taking New York City tourists for helicopter rides and also flying businessmen who preferred to commute to New York by helicopter.

The first French woman to receive her helicopter license was Valérie André. Unlike Reitsch and Shaw, whose careers centered on aviation, André was a brain surgeon. In 1949 she began to work as a surgeon in the French Army at war in Southeast Asia. Part of her duty was rescuing wounded French soldiers in the jungles, and André realized how much more easily she could perform such rescue missions if she had a helicopter. Accordingly she went back to France and worked for her license. When she returned to Hanoi, she saw to it that the Army issued her a helicopter. During her time in Indochina, she made over 100 helicopter flights and managed to rescue 165 soldiers and get them to hospitals.

ONE OF THE FOREMOST American women aviators of the 1950s was Geraldyn Cobb, who began to fly at the age of twelve. Like several other notable women aviators, Cobb's first teacher was her father. She first flew, in her father's plane, when she was still too small to reach the plane's controls without sitting on cushions. By the age of eighteen she had begun to earn her living as a pilot. Her flying jobs were varied: she dusted crops, taught flying and navigation, flew chartered planes, and served as sales manager and executive pilot of Aero Commander, Inc. One job Cobb did *not* hold was being a spy, although she once had a difficult time convincing some Ecuadorian officials of that when she made a crash landing there after flying out of Peru.

Cobb set many altitude and speed records in light planes in the fifties. In 1957 she soloed nonstop between Oklahoma and Guatamala, a distance of 1,504.7 miles; less than one month later she flew to a record altitude of 30,361 feet, which she afterwards improved by flying 37,000 feet. Some of her light plane records, in fact, beat those held by men pilots in the U.S.S.R.

In 1959 Jerrie Cobb received two special honors when she was named "Woman of the Year in Aviation" as well as the National Pilots Association's "Pilot of the Year." But Cobb gradually became less concerned with establishing a brilliant reputation than with using her flying for spiritual purposes. She is now a missionary pilot in South America.

The do-it-yourself days of aviation are not yet over. Recently Cobb sent this terse message from the Amazon Basin to fellow Ninety-Niner Betty Wright in Florida: *Engine main bearing failed at 80 hrs. Call Avco. Find Airplane. Home soonest. Jerrie.*[3] "For about thirty minutes I sat in disbelief," said Wright, before pulling herself together and arranging with Avco to have a new engine sent to Miami immediately. Cobb returned soon after that, "having hitch-hiked out of the jungle on a passing DC-3 loaded with two thousand empty beer bottles."[4] The two women finally managed to borrow a twin-engine Navajo from a third Ninety-Niner, Miriam Davis from Miami.

Wasting no time, the three women flew from Miami to Mitu, Colombia, where Cobb's plane was lying half-buried by mud. The plane was in fine shape, Wright reported, ex-

[3] Wright, Betty, "The Great Amazons Fly-In," *The 99 News* (Volume 6, Number 8, October, 1979), p. 16.

[4] Ibid., p. 16.

cept that every passerby "had felt called upon to engrave a symbol of love and care" on it. Cobb, Wright, and Davis spent "six hot, sticky, uncomfortable, sun-shining down, rainy days" installing the new engine themselves. Then Cobb flew off into the jungle again. Davis and Wright cooked a supper of Spam and creamed corn over a fire in a Tucano Indian village and set out for home.

TEST PILOTING has always been one of the most difficult areas for women to enter. One of the most successful pilots in this demanding field, however, is a woman, Marina Popovich of the Soviet Union. Popovich is also the holder of many world records, an honors graduate of a flying school whose directors found her such a skillful student that they made her an instructor, and the wife of Pavel Popovich, a famous cosmonaut—who once forbade her to fly.

Marina ignored even him, though, just as she had managed to ignore the medical examiners who told her late in the 1940s that she was too short by a quarter of an inch to join a flying club she had applied to. On hearing that news, Popovich joined a parachuting group instead and exercised so much that she managed somehow to add the necessary fraction to her height. She reapplied to the piloting club and this time was accepted.

After graduating from technical school, Popovich took a job at an aircraft plant and applied to a flying school. She was turned down again, and again she found a way to circumvent disappointment. She obtained a leave of absence from her job, moved to Moscow, and talked her way into a flying school there. It was there, after graduation, that she became a flight instructor.

Popovich had to stop flying for a time when her husband entered training for his 1962 space flight. She did so because he was so nervous about her flying that she feared it would keep him from concentrating on his training. Instead of flying, Marina entered an aviation institute. Thus equipped with more specialized knowledge, she began to work as a test pilot as soon as Pavel had finished his mission and she had graduated.

Probably the worst test flight for Popovich came when one plane's cabin suddenly depressurized at fifty-five thousand feet. Her reflexes must have been superb, because she was able to point the plane's nose up and turn on the emergency oxygen supply to save herself. Most other pilots would have fainted almost immediately. This sort of flight problem seems to have inconvenienced Popovich extremely rarely. She has broken at least thirteen world speed and distance records. Two of these events were especially welcome: the September 18, 1967, closed-circuit distance record that beat Jacqueline Cochran's record by 330 kilometers, and the first flight Popovich made in a fifty-ton Antaeus-22 turboprop. She had been frightened about flying such a huge plane, but on her first flight in it she and her crew set five world records, establishing a speed of 580 kilometers per hour on a two-thousand-kilometer route.

In private life, Popovich is just as tightly disciplined, though she has confessed that housework seems much harder to her than flying.

MOST PROFESSIONAL PILOTS love flying so much that they fly for recreation whenever they can. Gliding became one of the most popular forms of recreational flight after the war, and Hanna Reitsch's career as a glider pilot continued until her death in 1979. But the years just after the war were far from pleasant ones for her.

Towards the end of World War II, Reitsch spent her

days in the Alps organizing a resistance rescue movement. There she was captured by the Americans and imprisoned for months because she refused to go to the United States with them. During her stay in prison, Reitsch was almost kidnapped by Russians disguised as GI's; fortunately their attempt was unsuccessful. She was tried as a war criminal, but there was insufficient evidence to lead to her conviction, and Reitsch was allowed her freedom though not the right to fly. This was a bitter period in her life, made much more unhappy by the fact that her entire family had been lost in the war.

In 1952, when Germans were again permitted to fly, Reitsch took to the air once more. One of her first public appearances was at the 1952 international gliding meet in Spain. Her performance there was not very good because she had been earthbound for so long. But the same determination that had always carried her along helped Reitsch to improve until, three years later, she was once again Germany's champion glider.

In 1962 Ghanian President Kwame Nkrumah invited Reitsch to move to Ghana and establish a gliding school in Accra. Reitsch remained in Ghana for four years until Nkrumah—who had been her close friend—was overthrown in 1966. Reitsch then returned to Europe. She continued setting gliding records up until her death.

World War II helped make the glider popular with the public. Many gliders were used in the war, both for training military pilots and for antiaircraft target practice. Once the war was over, military gliders were sold extremely cheaply to the public, which probably did more than anything else to popularize the glider in this country.

The first national glider meet after the war took place in Texas in 1947. Virginia Bennis, the only woman of ninety-one United States glider pilots to hold an advanced gliding rating, set a new women's national distance record of ninety-four miles at the meet. Not long after that, Betty Loufek established a distance record of 124 miles. Her flight came to a terrifying, though harmless, end when a loose rope that had accidentally been stuck in the glider's wing caught and uprooted a cactus, brutally jerking the glider to a halt in midair. Then the craft floated docilely down to the ground.

British Anne Burns made one of the most famous glider flights in August, 1957, when she soared across the English Channel and became the first woman to have done so in a "freeglider" (untowed) flight. She also set a British women's speed record for a flight over a 200-kilometer triangular course. Burns averaged twenty-five miles per hour on that flight, a speed which seems unremarkable unless one realizes how difficult it is to fly a powerless aircraft over a set course.

THE FIRST WOMEN who dared to come down in parachutes would be terrified if they had to do it today. Skydiving has become an art, and among the most skillful practitioners of that art now are Soviet women. In 1975 Tatiana Morozycheva held third place in the world for number of jumps made by a woman—more than three thousand of them. She first became famous in 1967 when she set a world record in a group precision jump.

Morozycheva's favorite and most successful jump, though, is called the "30." To execute this jump, she must leave the plane at six thousand five hundred feet and perform an intricate series of maneuvers: a 360-degree horizontal turn, another 360-degree turn in a right spiral, and two somersaults. Her parachute must open before thirty seconds have elapsed from the start of the jump. Morozycheva explains her success at the "30" as springing from the fact that as a child she enjoyed both swimming and acrobatics—sports which honed her coordination and trained her to position her body precisely.

MARIE STIVERS IS HELPING to reintroduce one of the earliest forms of stunt-flying to the United States: wingwalking. She and her pilot call themselves the Flying Diamonds. They tour the country, paying for their biplane's fuel themselves and putting on shows wherever there is space enough and a big enough audience. Stivers says it feels as if she's performing "massive isometrics" when the plane dives. "It's difficult to breathe—we're both sustaining better than four g's." She finds it a most disconcerting experience. "As I'm looking down I see the ground. As we're going into the dive I look up and see the ground."

Women, Flight, & the Future

Thirty years ago . . . the man who talked of space ships and flights beyond the earth's atmosphere would have been considered likely material for an insane asylum.

Yet in the space of one generation we have seen many such dreams become actualities, taken for granted, and others are imminent.

So I have learned not to laugh off any prediction about the potentials of the air, however improbable. I consider it quite possible that during my lifetime interplanetary travel will become an accepted fact, and that ordinary citizens may plan vacations on the moon instead of at Miami Beach.

Of one thing I am certain—when space ships take off, I

shall be flying them, whether in my present bodily form or another.

—Ruth Nichols

THE SPACE AGE began on October 4, 1957. On that day, the Soviet Union launched the first man-made satellite, Sputnik I, into space. The United States, bristling with the spirit of competition, launched its first satellite, Explorer I, into the heavens on January 31, 1958. The space race had begun, and, from the women's point of view, it was won by the Russians on June 16, 1963, when Valentina Tereshkova entered near-Earth orbit in the Vostok 6. It would take fifteen years before the United States even began to choose women to enter space.

By the early 1960s it was well known that the Soviets were training women to become astronauts. At that time the relations between the United States and the U.S.S.R. were so strained that each country attempted to keep secret any news of its developments in space research. So America learned about women's taking part in the Soviet space program in a rather roundabout way. The West German Bochum Observatory, which, as one source said, was able to pick up the voices of Soviet cosmonauts, managed to "pick up" the sound of women's voices. The Russians also announced outright that they were considering using Mongolian women in their space program because Mongolians, living in mountainous areas, had become conditioned to an atmosphere with a lower-than-average level of oxygen.

Some Americans responded snidely to this news by answering that the Russians cared so little about Mongolians that they would be glad to sacrifice Mongolian lives if it would help Russia's space program surpass that of the United States. This was of course ridiculous, but in any case, Valentina Tereshkova was not Mongolian, and she made all forty-five of her historic orbits without any difficulty.

Tereshkova, a former factory worker, married a fellow cosmonaut—Andrian Nikolayev—soon after her space voyage. Though she continued to work in areas related to aerospace travel, preparing a doctoral dissertation at the Zhukovsky Air Force Engineering Academy, Tereshkova did not remain a cosmonaut. In 1968 she was elected president of the Soviet Women's Committee. The Committee is dedicated to promoting women's rights in all areas, to developing good relations with women's groups in other countries, and to working for peace among nations. In her capacity as president, Tereshkova helped to bring the Soviet Union's new constitution into being. Under the constitution, the state gives women equal access to education and training and also aids them with child care to make it easier for them to work.

Because American women have only very recently been chosen as actual astronauts, their role in the space program until now has necessarily been confined to jobs on earth. In 1960, in Albuquerque, New Mexico, Jerrie Cobb became the first woman to pass the tests for being an astronaut. These tests were just as tough as they had been for the male astronauts involved in Project Mercury. Among other things, Cobb underwent brainwave examinations and dozens of X-rays, drank radioactive water, kept her hands submerged in ice water for one minute, endured having cold water pumped into her ear canal, and swallowed several feet of hose for stomach tests. The result of Cobb's physical tests (and, later, the psychological ones) were superb, so much so that twenty-five women were asked to suffer through the same examination.

Janey Hart, a Whirly-Girl, also took and passed the physical tests for becoming an astronaut. So did identical twins Marion and Jan Dietrich, as well as Mary Wallace

Funk, who later became the first woman investigator for the National Transportation Safety Board. All of these tests took place in the early nineteen-sixties.

Carolyn Griner, a flight systems engineer, came closer than most American women to seeing what an actual orbit was like. Griner, along with three other women, participated in a five-day simulated space mission at the Marshall Space Flight Center. Griner has also worked in a pressurized space suit—underwater.

Jerrie Cobb frequently said that the perfect astronaut would be a "midget woman pilot from the Andes Mountains with a Ph.D." A candidate like this would be light enough to save NASA money (it costs many thousands of dollars per pound to put someone into orbit), small enough to save food costs, and well accustomed to living in a region with less oxygen than most of the world. Either no such candidate was ever found, or the United States was simply unwilling to put a woman into space. Although Jerrie Cobb was asked to serve as a NASA consultant, she and the rest of the women tested were never used as astronauts. Almost two decades passed after these tests before women began to be considered more seriously for the space program.

In January 1978, NASA officials selected thirty-five astronaut candidates—out of 8,079 applicants, 1,142 of them women—for its space shuttle program. It was the first group of candidates chosen in eight years. NASA announced that the thirty-five candidates were both highly qualified and highly motivated, and that unlike earlier applicants, many had hoped to become astronauts since childhood. They were distinguished for another reason as well, since three of the candidates were black, one was Japanese-American, and six were women: Dr. Shannon Lucid, a biochemist from Oklahoma; Dr. Anna L. Fisher, a physician from California; Dr. Judith A. Resnik, an electrical engineer in California; Sally K. Ride, a physics research assistant at

Stanford University; Dr. Margaret Rhea Seddon, a surgical resident in Memphis; and Kathryn D. Sullivan, a doctoral candidate in geology at Dalhousie University in Nova Scotia.

Two categories of astronauts were selected for the new space shuttle program—mission pilots, who were required to have experience as jet test pilots, and mission specialists. The training period took two years, with the first orbit scheduled for 1980. All six of the women candidates were picked to be mission specialists, which means that although they will not pilot the shuttle, they will have engineering, scientific, and medical duties and may also take part in extravehicular activities," or space walking.

The new space shuttle is designed to hold up to seven people in orbit. NASA officials estimate that the shuttle's first four test flights will be finished by 1982; manned flight missions are scheduled to start after that.

The six women are exceptionally qualified. The first American astronauts were primarily test pilots, while the candidates chosen as mission specialists in 1978 were primarily doctors or scientists. Among them the women have three Ph.D.'s, one master's degree, and two medical degrees. All of them gave up promising careers to become candidates, but none of them regret it. Some, like Sally Ride and Judith Resnik, feel that being an astronaut provides a logical extension to their careers. Ride, an astrophysicist, looks forward to performing experiments that can be done nowhere but in space. Resnik applied to become an astronaut because it looked clearly "like an opportunity for me to pursue what was most interesting to me in my career. It offered the diversity of working in a multidisciplinary field where I could have a turnover of projects." Becoming an astronaut, she says, "was strictly a career decision for me." Some of the women, on the other hand, thought being an astronaut was worth giving up a "regular job."

Ninety-Niner Margaret Rhea Seddon, for example, was

told by her sister that she was crazy to give up the chance of making five times more money as a surgeon than she could as an astronaut. But Seddon had already grown accustomed to doing the unexpected; few of her relatives had thought she should become a doctor in the first place. Born in 1947 in Murfreesboro, Tennessee, Seddon graduated from medical school in 1973 and spent three years in her surgical residency. During this time she was also able to put in about a hundred hours of private flying time. In fact—to his complete surprise—she asked her flying instructor to be one of the references on her astronaut's application.

Seddon once commented that becoming an astronaut will cost her more than any of the other five women because of what she might have earned as a surgeon. She added, though, that "this is what I want to do, so I'm going to do it. I feel the burden to succeed especially because of other women who want to go into the space program. They will be looking to us to do well so that NASA will accept more women in the future."[1] Seddon's special interest in nutrition led her to help plan the food systems for the orbital test flight.

Like all of the new mission specialists, Seddon's former career prepared her only partially for the work she will perform on the shuttle. Judith Resnik pointed out that "the object of mission-specialist training is to make us flexible enough to understand all sorts of disciplines that will be studied on the shuttle. Our chosen fields are important because they lead us into this field, and we maintain our proficiency in them, but our primary function is to be a mission specialist. Your training is useful in formulating how you solve problems, how you approach your work, how you interact with people on the job."

All the mission specialists trained together initially in

[1] Green, Janet, "Rhea Seddon, M.D.," *The 99 News*, p. 21.

1978; after the initial training period each was pretty much on his or her own. The physical training was diverse and strenuous, beginning with three days of "ocean survival exercises" in the 100-degree July weather of Homestead Air Force Base, Florida. Here the astronaut class practiced parachuting, bailing out, and managing to stay afloat, alone, in tiny plastic rafts on the open sea. (Sally Ride admitted later that doing the ocean survival exercise made her wonder why she was considered a smart person.)

Resnik added reassuringly, though, that the physical training was pretty much what she had expected. "There are a few unique things, obviously, that you have to go through—like being in a pressure suit or in a water facility or experiencing zero gravity in an airplane trajectory. But it isn't anything that requires any excessive amount of training."

Shannon Lucid, the only mother selected to be an astronaut, may have had more difficulty explaining to her three children that they couldn't go into space with her than in training. Since the Lucid family always traveled together, the children were certain that they would be allowed to accompany their mother.

Physicist Ann Whitaker, also a mother of a young child, was chosen, at approximately the same time as the six women candidates, as one of the six American finalists to be a participant in the joint U.S.–European Spacelab mission scheduled for 1980. She was the only woman picked for the mission.

Dr. Anna Fisher, the only other of the six women space shuttle candidates who is married, found, of course, that becoming an astronaut meant that she saw little of her husband—especially because his job as an emergency-room physician kept him busy as well. With the other trainees, Fisher put in twelve-hour days at the NASA base in Houston, where she studied astronomy, celestial navigation, and space walking. In what there was of her spare time, she jogged four miles every other day, lifted weights, played racquetball, made her first solo flight, and went skiing and canoeing with her husband. She also shared the housework with him and, in her second year of astronaut training, returned to practicing emergency-room medicine to polish her skill in that area. (As a little girl, Fisher had already decided that when she grew up there would be space stations that would need doctors.)

The Fishers both applied to become astronauts at the same time, and they learned on their honeymoon that Anna had been chosen. William Fisher was turned down four months later. He spent the next two years studying engineering and also received his private pilot's license so that NASA officials might choose him later. In June, 1980, the Fishers became the country's first astronaut couple when William too was chosen as one of nineteen new candidates for the space shuttle project. Anna was especially delighted by the news that he would begin training because it meant that she would no longer have to feel guilty about describing how much fun she had been having.

It is unlikely that the Fishers will be sent into orbit together. Their skills are so similar that there would be no point in assigning them to the same shuttle flight. Both expect to conduct some experiments in orbit and to practice medicine, perhaps emergency medicine if that is necessary.

Anna Fisher reported that she found less discrimination as an astronaut at NASA than she had met with as a doctor. "The only trouble the women astronaut candidates have had is getting clothing to fit." NASA officials have tried wherever possible to keep sex discrimination out of the training program, although they have changed the design of the women's pressurized suits somewhat. Norris expected that lack of space aboard the shuttle will not create problems between the sexes, as one male astronaut predicted there would be. This is simply because the shuttle is much larger than the Gemini or Apollo spacecraft. Anyway, said Judith Resnik, "I don't think that the amount of space, or lack of it, has anything to do with whether a man or a woman is fit to participate."

The fact that the women astronaut candidates have met with such high official and public approval is a sign of how greatly social attitudes have changed since the beginning of the space age. Just as welcome a change may be seen in the fact that the six new astronauts think of their Russian counterparts as colleagues, not competitors. Although these six American women will not be sent off into space for several years to come, there is no doubt that they have already succeeded.

Bibliography

Adams, Jean, and Kimball, Margaret. *Heroines of the Sky*. New York: Doubleday & Co., Inc., 1942.

Arnold, Maj. Terry A. "Baptizing the New Breed." *Airman*. October, 1977.

Auriol, Jacqueline. *Vivre et Voler*. Paris: Éditions Flammarion, 1968.

Bacon, Gertrude. *Memories of Land and Sky*. London: Methuen & Co., Ltd., 1928.

Cochran, Jacqueline. *The Stars at Noon*. Boston: Little, Brown & Company, 1954.

Collins, Helen F. "From Plane Captains to Pilots." *Naval Aviation News*. July, 1977.

Crane, Mardo. "The Women with Silver Wings." *The 99 News*. Volume 5, Number 8, 1978.

Dwiggins, Don. *Riders of the Wind: The Story of Ballooning*. New York: Hawthorn Books, Inc., 1973.

Earhart, Amelia. *The Fun of It*. Chicago: The Academy Press, Ltd., 1977.

Gibbs-Smith, Charles Harvard. *Flight Through the Ages*. New York: Thomas Y. Crowell Co., 1974.

Glines, Lt. Colonel. *Lighter-than-Air Flight*. New York: Franklin Watts, Inc., 1965.

Green, Janet. "Rhea Seddon, M.D." *The 99 News*. Volume 5, Number 8, 1978.

Hillinger, Charles. "Women Pilots Share Achievements." The Los Angeles *Times*. May 16, 1980.

Horn, Maurice, ed. *The World Encyclopedia of Comics*. New York: Chelsea House Publishers, 1976.

Jablonski, Edward. *Ladybirds: Women in Aviation*. New York: Hawthorn Books, Inc., 1968.

Josephy, Alvin, Jr., ed. *The American Heritage History of Flight*. New York: American Heritage Publishing Co., Inc., 1962.

Lauwick, Hervé. *Heroines of the Sky*. New Rochelle, N.Y.: Soccer Associates, 1961.

May, Charles Paul. *Women in Aeronautics*. New York: Thomas Nelson and Sons, 1962.

Nichols, Ruth. *Wings for Life*. New York: J. B. Lippincott Company, 1957.

Nye, Sandy. "Up Front with Judy." *Naval Aviation News*. July, 1977.

Owen, David. *Flight*. New York: Harmony Books, 1978.

Planck, Charles E. *Women with Wings*. New York: Harper & Brothers, 1942.

Reitsch, Hanna. *Flying Is My Life*. New York: Van Rees Press, 1954.

Roseberry, C. R. *The Challenging Skies*. New York: Doubleday & Co., Inc., 1966.

Smith, Linell. "In Control." The Baltimore *Evening Sun*. August 15, 1979.

Thaden, Louise. *High, Wide and Frightened.* New York: Stackpole Sons, 1938.

Van Wagenen Keil, Sally. *Those Wonderful Women in Their Flying Machines.* New York: Rawson, Wade Publishers, Inc., 1979.

Vecsey, George, and Dade, George C. *Getting Off the Ground.* New York: E. P. Dutton, 1979.

Wright, Betty. "The Great Amazons Fly-in." *The 99 News.* Volume 6, Number 8. October, 1979.

Index

Index

Index